Bernhard Bundschuh
Eric Christian Nana Wadjounnie

The Doppler Effect in Terms of System Theory

Bernhard Bundschuh
Eric Christian Nana Wadjounnie

The Doppler Effect in Terms of System Theory

Relativistic and non-relativistic effect

Südwestdeutscher Verlag für Hochschulschriften

Impressum/Imprint (nur für Deutschland/ only for Germany)
Bibliografische Information der Deutschen Nationalbibliothek: Die Deutsche Nationalbibliothek verzeichnet diese Publikation in der Deutschen Nationalbibliografie; detaillierte bibliografische Daten sind im Internet über http://dnb.d-nb.de abrufbar.
Alle in diesem Buch genannten Marken und Produktnamen unterliegen warenzeichen-, marken- oder patentrechtlichem Schutz bzw. sind Warenzeichen oder eingetragene Warenzeichen der jeweiligen Inhaber. Die Wiedergabe von Marken, Produktnamen, Gebrauchsnamen, Handelsnamen, Warenbezeichnungen u.s.w. in diesem Werk berechtigt auch ohne besondere Kennzeichnung nicht zu der Annahme, dass solche Namen im Sinne der Warenzeichen- und Markenschutzgesetzgebung als frei zu betrachten wären und daher von jedermann benutzt werden dürften.

Verlag: Südwestdeutscher Verlag für Hochschulschriften Aktiengesellschaft & Co. KG
Dudweiler Landstr. 99, 66123 Saarbrücken, Deutschland
Telefon +49 681 37 20 271-1, Telefax +49 681 37 20 271-0, Email: info@svh-verlag.de
Zugl.: Forschungsarbeit

Herstellung in Deutschland:
Schaltungsdienst Lange o.H.G., Berlin
Books on Demand GmbH, Norderstedt
Reha GmbH, Saarbrücken
Amazon Distribution GmbH, Leipzig
ISBN: 978-3-8381-0649-6

Imprint (only for USA, GB)
Bibliographic information published by the Deutsche Nationalbibliothek: The Deutsche Nationalbibliothek lists this publication in the Deutsche Nationalbibliografie; detailed bibliographic data are available in the Internet at http://dnb.d-nb.de.
Any brand names and product names mentioned in this book are subject to trademark, brand or patent protection and are trademarks or registered trademarks of their respective holders. The use of brand names, product names, common names, trade names, product descriptions etc. even without a particular marking in this works is in no way to be construed to mean that such names may be regarded as unrestricted in respect of trademark and brand protection legislation and could thus be used by anyone.

Publisher:
Südwestdeutscher Verlag für Hochschulschriften Aktiengesellschaft & Co. KG
Dudweiler Landstr. 99, 66123 Saarbrücken, Germany
Phone +49 681 37 20 271-1, Fax +49 681 37 20 271-0, Email: info@svh-verlag.de

Copyright © 2009 by the author and Südwestdeutscher Verlag für Hochschulschriften Aktiengesellschaft & Co. KG and licensors
All rights reserved. Saarbrücken 2009

Printed in the U.S.A.
Printed in the U.K. by (see last page)
ISBN: 978-3-8381-0649-6

Preface

I am teaching system theory for 15 years now at the University of Applied Sciences in Merseburg, Germany. To many students system theory appears to be an incomprehensible subject far from practical application and fraught with complicated mathematical formulas. In electrical engineering system theory is mainly used for modelling of communication systems and automatic control systems. Due to the widespread use of digital signal processing more and more practical applications become economically feasible.

A less widespread application of system theory is the mathematical modelling of physical phenomena e. g. in optics and acoustics. The Doppler Effect is one of the most well known physical phenomena. There are plenty of applications in various fields of science and technology. It is even a very popular subject for high school projects. At first glance it appears to be quite unlikely to find new aspects of the Doppler Effect. At least Christian Doppler's famous paper "Über das farbige Licht der Doppelsterne" was published as early as 1842.

However in 1999 I had the idea to derive a generalisation of the Doppler Effect to non-sinusoidal signals and to arbitrary trajectories of transmitters and/or receivers using methods of system theory.

Soon after some investigations had led to promising results I published a conference paper. However this paper was far from covering the subject in all aspects. The relativistic effect was not included at all and only a special case of radar – reflector configurations was considered. Due to other activities I did not pursue the subject until 2007. Then Eric Christian Nana Wadjounnie looked for a diploma thesis project which should involve plenty of mathematics, a rather unusual request from a student. Soon after that we started to extend the coverage of the derivation of the Doppler Effect in terms of system theory. It turned out quickly that my approach from 1999 was only correct in the special case of a stationary receiver. After the extension to the relativistic Doppler Effect our approach was confirmed by comparison with some results in Einstein's famous paper on special relativity from 1905.

At the end of the day we managed to develop a very general derivation in terms of system theory. A quick look to this book shows that the required amount of mathematics is quite involved. On the other hand we gained profound insight into the generation of the Doppler Effect. Most of the envisaged generalisations could be realised. However the coverage of the subject can be extended in several directions. Such there may be enough sub-topics e. g. for some more students who may look for thesis projects which should involve plenty of mathematics. Perhaps in a couple of years the publication of another book with extended coverage of the subject will be possible.

Bernhard Bundschuh   Merseburg, April 2009

Acknowledgement

The authors would like to thank Monika Trundt and Jonker Teske for carefully checking the English translation.

Contents

1 Introduction 9

2 Underlying System Theory 10

3 Uniform Motion of Transmitter and Receiver on a Line 11

   3.1 Non-relativistic Doppler Effect 11

      3.1.1 Calculation of the received Signal 13
      3.1.2 Auto Correlation Function of the received Signal 15
      3.1.3 Fourier Transform of the received Signal 16

   3.2 Relativistic Doppler Effect 17

      3.2.1 Stationary Receiver and uniform Motion of Transmitter 17
      3.2.2 Stationary Transmitter and uniform Motion of Receiver 18
      3.2.3 Uniform Motion of Transmitter and Receiver 19

4 Uniform Motion of Transmitter and Receiver in Space 23

   4.1 Non-relativistic Doppler Effect 23

      4.1.1 Calculation of the received Signal in General Form 24
      4.1.2 Stationary Receiver 26
      4.1.3 Stationary Transmitter 28
      4.1.4 Auto Correlation Function of the received Signal 30
      4.1.5 Fourier Transform of the received Signal 32

   4.2 Relativistic Doppler Effect 33

      4.2.1 Stationary Receiver 33
      4.2.2 Stationary Transmitter 37
      4.2.3 Motion of Transmitter and Receiver 38

5 Monostatic Radar 40

   5.1 Stationary Radar and uniform Motion of Reflector 40

      5.1.1 Non-relativistic Doppler Effect 40
      5.1.2 Relativistic Doppler Effect 45

   5.2 Stationary Reflector and uniform Motion of Radar 47

   5.3 Airborne Radar over Ground 49

      5.3.1 Calculation of the received Signal in General Form 50
      5.3.2 Application to Synthetic Aperture Radar 52

6	**Non-uniform Motion**	**55**
6.1	Stationary Receiver and non-uniform Motion of Transmitter	55
6.1.1	Non-relativistic Doppler Effect	55
6.1.2	Relativistic Doppler Effect	57
6.2	Stationary Transmitter and non-uniform Motion of Receiver	58
6.2.1	Non-relativistic Doppler Effect	59
6.2.2	Relativistic Doppler Effect	59
6.3	Experimental Test	60
7	**Conclusions**	**62**
A	**Appendices**	**63**
A.1	Derivation of Equation 4.24	63
A.2	Relativistic Subtraction of Velocity Vectors	64
A.3	Derivation of Equation 5.9	67
A.4	Derivation of Equation 5.15a	68
A.5	Derivation of Equation 5.25	70
A.6	Derivation of Equations 5.33a and 5.33b	71
A.7	Taylor Expansion for Synthetic Aperture Radar	74
Acronyms and Symbols		**83**
References		**85**
Index		**86**

1   Introduction

In 1842 Christian Doppler's famous paper "Über das farbige Licht der Doppelsterne" was published in Prague [1]. The generalisation covering the relativistic Doppler Effect was published 1905 by Albert Einstein in his also very famous paper "Zur Elektrodynamik bewegter Körper" [2]. Now, more than 100 years later, the question may arise: Are there still new insights into the Doppler Effect to be gained?

In Physics textbooks the modification of frequency caused by the non-relativistic Doppler Effect is usually expressed as in eq. 1.1.

$$f_r = f_t \cdot \frac{1 - v_r/c_0}{1 - v_t/c_0} \qquad (1.1)$$

v_t and v_r are the velocities of the transmitter resp. the receiver. f_t and f_r are the frequencies of the transmitted resp. the received signal. c_0 is the velocity of propagation of the signal. In the non-relativistic case c_0 is the velocity of sound. In the relativistic case it is the velocity of light.

Since there is no privileged reference frame in special relativity the relativistic Doppler Effect is determined only by the relative velocity (relativistic subtraction of velocities) v_d of the transmitter and the receiver.

$$f_r = f_t \sqrt{\frac{1 - v_d/c_0}{1 + v_d/c_0}} \qquad (1.2)$$

Both equations are only valid if some basic assumptions are met. The transmitter and the receiver move with constant velocities on the same line in space. The transmitted signal is a sinusoid of frequency f_t.

The derivation in terms of system theory enables a more general description of the Doppler Effect. However the required amount of mathematics is much higher than in eqs. 1.1 and 1.2. Fig. 1 schematically illustrates the generalisations that can be gained from the new approach.

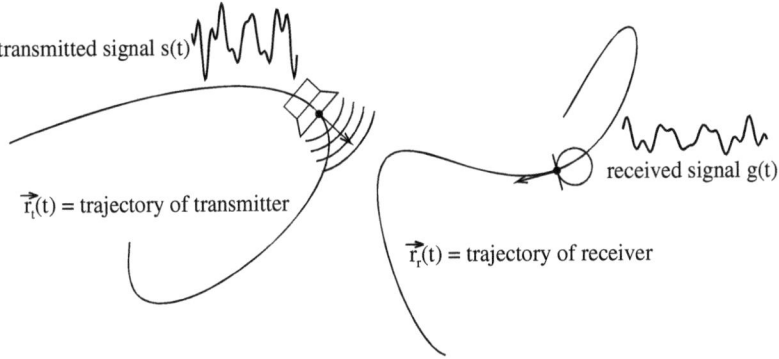

Fig. 1: Transmitter and receiver on arbitrary trajectories with arbitrary waveforms

2 Underlying System Theory

If a stationary transmitter and a stationary receiver are separated by distance d in space, the travel time of the signal from the transmitter to the receiver is d/c_0.

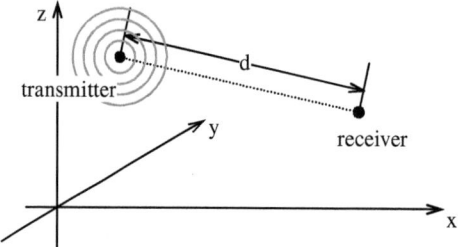

Fig. 2: Stationary transmitter and stationary receiver

In fig. 1 s(t) is the transmitted signal. g(t) is the received signal. Using convolution and the shifting property of the delta function $\delta(t)$ [3] g(t) is easily calculated according to eq. 2.1.

$$g(t) = s(t) * \delta\left(t - \frac{d}{c_0}\right) = \int_{-\infty}^{\infty} s(\tau) \underbrace{\delta\left(t - \tau - \frac{d}{c_0}\right)}_{h(t-\tau)} d\tau = s\left(t - \frac{d}{c_0}\right) \qquad (2.1)$$

Mathematically the signal delay is formulated as the convolution of the transmitted signal and the impulse response $h(t) = \delta(t-d/c_0)$. In reality there is also some signal attenuation depending on the distance of transmitter and receiver as well as on the orientation of transmitter and receiver in space etc.. This kind of attenuation is not considered throughout this book.

The Doppler Effect arises if the transmitter and the receiver move relative to each other, in other words, the distance is no longer constant but depends on time. The convolution in eq. 2.1 must now be replaced by a more general linear superposition using a linear time-variant impulse response $h(t,\tau)$ according to eq. 2.2.

$$g(t) = \int_{-\infty}^{\infty} s(\tau) \underbrace{\delta\left(t - \tau - d(\tau)/c_0\right)}_{h(t,\tau)} d\tau \qquad (2.2)$$

3   Uniform Motion of Transmitter and Receiver on a Line

Fig. 3 schematically shows a transmitter in motion and a stationary receiver. If the transmitter approaches the receiver, the frequency of a received sinusoidal waveform increases. If the transmitter moves away from the receiver, the frequency decreases.

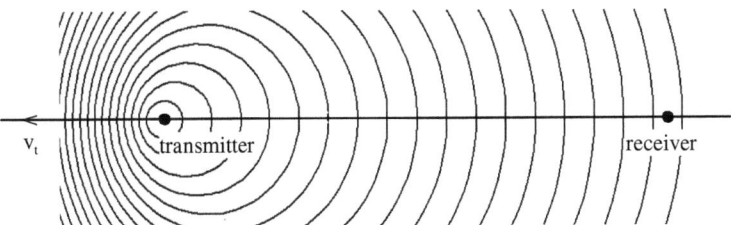

Fig. 3: Emission of a signal by a moving transmitter

3.1   Non-relativistic Doppler Effect

Without loss of generality the coordinate system is defined such that the transmitter and the receiver move with uniform velocities on the x-axis. The locations of the transmitter resp. the receiver are functions of time according to:

$$x = x_{t0} + v_t t \quad \text{resp.} \quad x = x_{r0} + v_r t \tag{3.1}$$

x_{t0} resp. x_{r0} are the locations of the transmitter resp. the receiver at time $t = 0$.

At time τ the transmitter emits a spherical wave. Eq. 3.2 characterises the evolution of the sphere as a function of time.

$$\left(x - (x_{t0} + v_t \tau)\right)^2 = c_0^2 (t - \tau)^2 \tag{3.2}$$

The squared radius of the sphere at time t is on the right-hand side of the equation. The radius is proportional to the velocity of propagation c_0 of the signal.

Fig. 4 illustrates the situation.

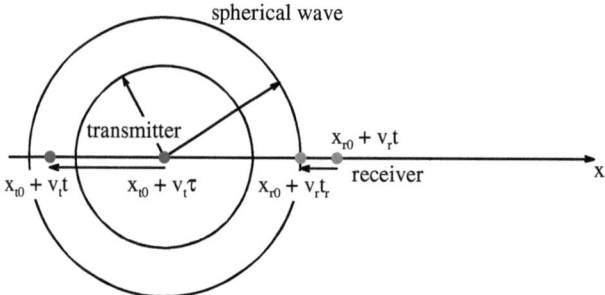

Fig. 4: Propagation of a spherical wave from the transmitter to the receiver

At time t_r the spherical wave arrives at the receiver. This time is calculated using the trajectory of the receiver according to eq. 3.1 in eq. 3.2.

$$(x_{r0} + v_r t_r - x_{t0} - v_t \tau)^2 = c_0^2 (t_r - \tau)^2 \qquad (3.3)$$

There are two solutions for t_r.

$$\pm(x_{r0} + v_r t_r - x_{t0} - v_t \tau) = c_0 (t_r - \tau) \qquad (3.4a)$$

$$\pm(x_{r0} - x_{t0}) \pm (v_r t_r - v_t \tau) = c_0 t_r - c_0 \tau \Rightarrow (c_0 \mp v_r) t_r = (c_0 \mp v_t) \tau \pm (x_{r0} - x_{t0})$$

$$t_r = \frac{c_0 \mp v_t}{c_0 \mp v_r} \tau \pm \frac{x_{r0} - x_{t0}}{c_0 \mp v_r} \qquad (3.4b)$$

The correct solution is determined by the causality condition $t_r \geq \tau$. The space-time diagram in fig. 5 illustrates the situation.

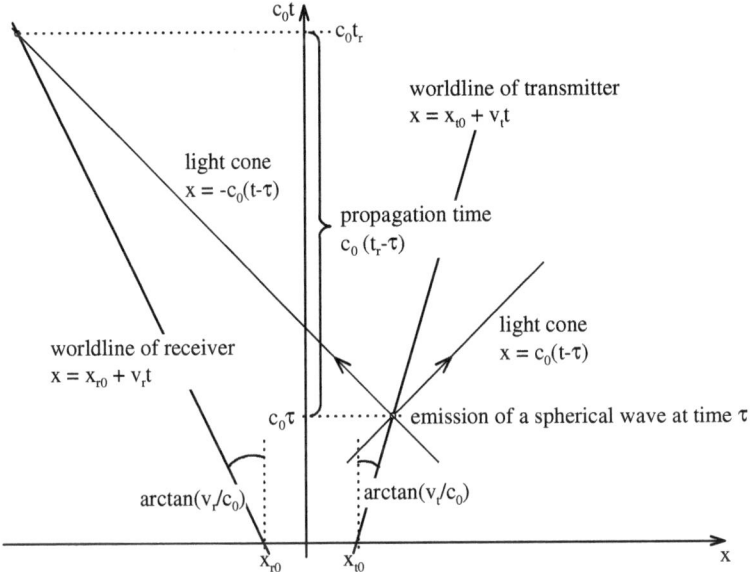

Fig. 5: Space-time diagram

Strictly speaking, the term light cone is only correct for the relativistic Doppler Effect. But here it is also used to characterise the propagation of acoustic signals. The slopes of the branches of the light cone are +1 resp. -1. The point of intersection of the light cone and the worldline of the receiver determines the above mentioned time t_r. The causality condition $t_r \geq \tau$ determines the valid branch of the light cone.

3.1.1 Calculation of the received Signal

Case a
$$t_r(\tau) = \frac{c_0 - v_t}{c_0 - v_r}\tau + \frac{x_{r0} - x_{t0}}{c_0 - v_r} = \frac{1 - v_t/c_0}{1 - v_r/c_0}\tau + \frac{1}{1 - v_r/c_0}\frac{x_{r0} - x_{t0}}{c_0} \quad (3.5a)$$

The time of propagation $t_r(\tau) - \tau$ from the transmitter to the receiver is used in the linear superposition according to eq. 2.2.

$$g(t) = \int_{-\infty}^{\infty} s(\tau)\delta\big(t - \tau - (t_r(\tau) - \tau)\big)d\tau = \int_{-\infty}^{\infty} s(\tau)\delta\big(t - t_r(\tau)\big)d\tau \quad (3.6)$$

$t_r(\tau)$ according to eq. 3.5a is used in eq. 3.6 followed by a change of variables according to eq. 3.7a.

The integral is calculated using the sifting property of the delta function [3].

$$g(t) = \int_{-\infty}^{\infty} s(\tau) \delta(t - t_r(\tau)) d\tau = \int_{-\infty}^{\infty} s(\tau) \delta\left(\underbrace{t - \left(\frac{1 - v_t/c_0}{1 - v_r/c_0} \tau + \frac{1}{1 - v_r/c_0} \frac{x_{r0} - x_{t0}}{c_0}\right)}_{w}\right) d\tau \quad (3.7a)$$

The delta function in eq. 3.7a is a particular example of the general time-variant impulse response $h(t,\tau)$ according to eq. 2.2. The received signal is obtained using some straightforward calculations.

$$\tau = \frac{1 - v_r/c_0}{1 - v_t/c_0}\left(t - w - \frac{1}{1 - v_r/c_0}\frac{x_{r0} - x_{t0}}{c_0}\right) \Rightarrow \frac{d\tau}{dw} = -\frac{1 - v_r/c_0}{1 - v_t/c_0}$$

$$g(t) = -\frac{1 - v_r/c_0}{1 - v_t/c_0} \int_{\infty}^{-\infty} s\left(\frac{1 - v_r/c_0}{1 - v_t/c_0}\left(t - w - \frac{1}{1 - v_r/c_0}\frac{x_{r0} - x_{t0}}{c_0}\right)\right) \delta(w) dw$$

$$g(t) = \left|\frac{1 - v_r/c_0}{1 - v_t/c_0}\right| \cdot s\left(\frac{1 - v_r/c_0}{1 - v_t/c_0}\left(t - \frac{1}{1 - v_r/c_0}\frac{x_{r0} - x_{t0}}{c_0}\right)\right) \quad (3.8a)$$

Three effects determine the received signal: time scaling, amplitude scaling and time delay. The time and amplitude scaling factors are functions of the velocities of the transmitter <u>and</u> the receiver. The time delay is a function of the velocity of the receiver only. Both velocities are normalised to the velocity of propagation of the signal. Only if the transmitted signal s(t) is a sinusoidal waveform time scaling is equivalent to a frequency shift according to eq. 1.1. Thus the term Doppler Scaling would be more general than the commonly used term Doppler Shift.

If the velocity of the transmitter is equal to the velocity of sound c_0, the amplitude of the received signal becomes infinite. This is equivalent to the supersonic boom caused by an aircraft going supersonic.

If the velocity of the receiver is equal to the velocity of sound, the amplitude of the received signal becomes zero and the time delay becomes infinite. In this case the signal never reaches the receiver.

If the transmitter ($v_t = 0$) resp. the receiver ($v_r = 0$) is stationary there are two special cases.

$$g(t) = \left|1 - \frac{v_r}{c_0}\right| \cdot s\left(\left(1 - \frac{v_r}{c_0}\right)\left(t - \frac{1}{1 - v_r/c_0}\frac{x_{r0} - x_{t0}}{c_0}\right)\right) \quad (3.9)$$

$$g(t) = \left|\frac{1}{1 - v_t/c_0}\right| \cdot s\left(\frac{1}{1 - v_t/c_0}\left(t - \frac{x_{r0} - x_{t0}}{c_0}\right)\right) \quad (3.10)$$

Case b
$$t_r(\tau) = \frac{c_0 + v_t}{c_0 + v_r}\tau - \frac{x_{r0} - x_{t0}}{c_0 + v_r} = \frac{1 + v_t/c_0}{1 + v_r/c_0}\tau - \frac{1}{1 + v_r/c_0}\frac{x_{r0} - x_{t0}}{c_0} \quad (3.5b)$$

The received signal is now calculated using the change of variables according to eq. 3.7b. The integral is again calculated using the sifting property of the delta function [3].

$$g(t) = \int_{-\infty}^{\infty} s(\tau)\delta(t - t_r(\tau))d\tau = \int_{-\infty}^{\infty} s(\tau)\delta\underbrace{\left(t - \left(\frac{1 + v_t/c_0}{1 + v_r/c_0}\tau - \frac{1}{1 + v_r/c_0}\frac{x_{r0} - x_{t0}}{c_0}\right)\right)}_{w}d\tau \quad (3.7b)$$

The received signal is obtained using some straightforward calculations.

$$\tau = \frac{1 + v_r/c_0}{1 + v_t/c_0}\left(t - w + \frac{1}{1 + v_r/c_0}\frac{x_{r0} - x_{t0}}{c_0}\right) \Rightarrow \frac{d\tau}{dw} = -\frac{1 + v_r/c_0}{1 + v_t/c_0}$$

$$g(t) = -\frac{1 + v_r/c_0}{1 + v_t/c_0}\int_{\infty}^{-\infty} s\left(\frac{1 + v_r/c_0}{1 + v_t/c_0}\left(t - w + \frac{1}{1 + v_r/c_0}\frac{x_{r0} - x_{t0}}{c_0}\right)\right)\delta(w)dw$$

$$g(t) = \left|\frac{1 + v_r/c_0}{1 + v_t/c_0}\right| \cdot s\left(\frac{1 + v_r/c_0}{1 + v_t/c_0}\left(t + \frac{1}{1 + v_r/c_0}\frac{x_{r0} - x_{t0}}{c_0}\right)\right) \quad (3.8b)$$

The discussion of this result and the special cases is equivalent to case a.

3.1.2 Auto Correlation Function (ACF) of the received Signal (only case a exemplarily)

The calculation of the ACF $\varphi_{gg}(\Delta t)$ of the received signal $g(t)$ according to eq. 3.7a is demonstrated without loss of generality for the special case of a stationary receiver ($v_r = 0$) located at $x_{r0} = 0$.

$$\varphi_{gg}(\Delta t) = \int_{-\infty}^{\infty} g(t)g(t - \Delta t)dt \quad (3.11)$$

Using two subsequent changes of variables the ACF of the received signal can be expressed as a function of the ACF $\varphi_{ss}(\Delta t)$ of the transmitted signal $s(t)$.

First change of variables:

$$\varphi_{gg}(\Delta t) = \int_{-\infty}^{\infty}\int_{-\infty}^{\infty} s(\tau)\delta\left(t - \left(1 - \frac{v_t}{c_0}\right)\tau + \frac{x_{t0}}{c_0}\right)d\tau \cdot \int_{-\infty}^{\infty} s(\hat{\tau})\delta\underbrace{\left(t - \Delta t - \left(1 - \frac{v_t}{c_0}\right)\hat{\tau} + \frac{x_{t0}}{c_0}\right)}_{u}d\hat{\tau}\,dt$$

$$t = u + \Delta t + \left(1 - \frac{v_t}{c_0}\right)\hat{\tau} - \frac{x_{t0}}{c_0} \Rightarrow dt = du$$

Second change of variables:

$$\varphi_{gg}(\Delta t) = \int_{-\infty}^{\infty}\int_{-\infty}^{\infty}\int_{-\infty}^{\infty} s(\tau)s(\hat{\tau})\delta\left(\underbrace{u+\Delta t+\left(1-\frac{v_t}{c_0}\right)\hat{\tau} - \frac{x_{t0}}{c_0} - \left(1-\frac{v_t}{c_0}\right)\tau + \frac{x_{t0}}{c_0}}_{w}\right)\delta(u)\,du\,d\hat{\tau}\,d\tau$$

$$\left(1-\frac{v_t}{c_0}\right)\hat{\tau} = w - \Delta t + \left(1-\frac{v_t}{c_0}\right)\tau \Rightarrow \hat{\tau} = \tau - \frac{\Delta t - w}{1-v_t/c_0} \Rightarrow \frac{d\hat{\tau}}{dw} = \frac{1}{1-v_t/c_0}$$

Now the sifting property of the delta function [3] is applied twice.

$$\varphi_{gg}(\Delta t) = \left|\frac{1}{1-v_t/c_0}\right| \underbrace{\int_{-\infty}^{\infty}\int_{-\infty}^{\infty} s(\tau)s\left(\tau - \frac{\Delta t - w}{1-v_t/c_0}\right)d\tau}_{\varphi_{ss}\left(\frac{\Delta t - w}{1-v_t/c_0}\right)} \underbrace{\int_{-\infty}^{\infty}\delta(u+w)\delta(u)\,du}_{\delta(w)}\,dw$$

$$\varphi_{gg}(\Delta t) = \left|\frac{1}{1-v_t/c_0}\right| \int_{-\infty}^{\infty} \varphi_{ss}\left(\frac{\Delta t - w}{1-v_t/c_0}\right)\delta(w)\,dw = \left|\frac{1}{1-v_t/c_0}\right| \varphi_{ss}\left(\frac{\Delta t}{1-v_t/c_0}\right) \quad (3.12)$$

The ACF of the received signal is equivalent to the time and amplitude scaled ACF of the transmitted signal. Therefore the derivation of the Doppler Effect in terms of system theory is also applicable to stochastic signals, e. g. noise.

3.1.3 Fourier Transform of the received Signal (only case a exemplarily)

The calculation of the Fourier transform $\underline{G}(f)$ of the received signal g(t) according to eq. 3.7a is demonstrated without loss of generality for the special case of a stationary receiver ($v_r = 0$) located at $x_{r0} = 0$.

$$\underline{G}(f) = \int_{-\infty}^{\infty} g(t)e^{-j2\pi ft}\,dt = \left|\frac{1}{1-v_t/c_0}\right| \int_{-\infty}^{\infty} s\left(\frac{1}{1-v_t/c_0}\left(t + \frac{x_{t0}}{c_0}\right)\right) e^{-j2\pi ft}\,dt \quad (3.13)$$

Change of variables $\quad \frac{1}{1-v_t/c_0}\left(t + \frac{x_{t0}}{c_0}\right) = u \Rightarrow t = \left(1-\frac{v_t}{c_0}\right)u - \frac{x_{t0}}{c_0} \Rightarrow \frac{dt}{du} = 1 - \frac{v_t}{c_0}$

$$\underline{G}(f) = \left|\frac{1}{1-v_t/c_0}\right| \int_{-\infty}^{\infty} s(u) e^{-j2\pi f\left(\left(1-\frac{v_t}{c_0}\right)u - \frac{x_{t0}}{c_0}\right)} \left(1-\frac{v_t}{c_0}\right) du = \int_{-\infty}^{\infty} s(u) e^{-j2\pi\left(1-\frac{v_t}{c_0}\right)fu} e^{j2\pi f\frac{x_{t0}}{c_0}}\,du$$

$$\underline{G}(f) = \underline{S}\left(\left(1-\frac{v_t}{c_0}\right)f\right) \cdot e^{j2\pi f\frac{x_{t0}}{c_0}} \quad (3.14)$$

Time scaling of the signal causes reciprocal frequency scaling. This is known as scaling property of the Fourier transform [3]. Time delay causes a linear phase shift.

3.2 Relativistic Doppler Effect

The transition from the non-relativistic Doppler Effect to the relativistic one is accomplished by application of the Lorentz transform [4].

3.2.1 Stationary Receiver and uniform Motion of Transmitter

Eq. 3.5a is the starting point for the relativistic calculation. The velocity v_r of the receiver is set to zero. An equivalent calculation could be performed starting from eq. 3.5b. Since no new insight could be gained from this calculation it is omitted.

First the coordinate system is shifted such that the receiver is located at x = 0. Fig. 6 illustrates this modification. The signals are not affected by this shifting operation. Without this operation the time delay of the received signal would be a function of the absolute position x_{r0}. Since there is no privileged reference frame in special relativity, the time delay must be a function of the difference $x_{r0} - x_{t0}$ only.

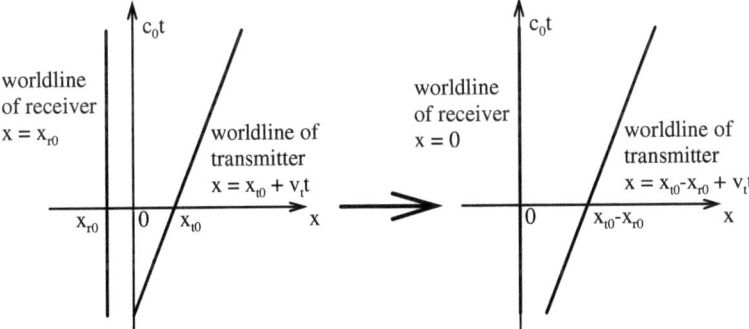

Fig. 6: Shifting of the space-time-diagram

The Lorentz transform [4] is applied to the receiving time t_r of the signal. Shifting of the coordinate system followed by a Lorentz transform is equivalent to a purely temporal Lorentz transform according to eq. 3.15 without shifting of the coordinate system. In the following only this simplified Lorentz transform is used.

$$t' = \frac{t}{\sqrt{1 - v_t^2/c_0^2}} \quad \text{resp.} \quad t = t' \cdot \sqrt{1 - v_t^2/c_0^2} \tag{3.15}$$

Application of eq. 3.15 to t_r according to eq. 3.5a using $v_r = 0$ results in:

$$t'_r(\tau) = \frac{t_r(\tau)}{\sqrt{1 - v_t^2/c_0^2}} = \frac{1}{\sqrt{1 - v_t^2/c_0^2}} \left(\left(1 - \frac{v_t}{c_0}\right)\tau + \frac{x_{r0} - x_{t0}}{c_0} \right) \tag{3.16}$$

The calculation of the received signal is equivalent to the non-relativistic case.

$$g(t) = \int_{-\infty}^{\infty} s(\tau)\delta(t - t'_r(\tau))d\tau = \int_{-\infty}^{\infty} s(\tau)\delta\underbrace{\left(t - \left(\frac{1-v_t/c_0}{\sqrt{1-v_t^2/c_0^2}}\tau + \frac{1}{\sqrt{1-v_t^2/c_0^2}}\frac{x_{r0}-x_{t0}}{c_0}\right)\right)}_{w}d\tau$$

A change of variables is followed by an application of the sifting property of the delta function.

$$\tau = \sqrt{\frac{1+v_t/c_0}{1-v_t/c_0}}\left(t - w - \frac{1}{\sqrt{1-v_t^2/c_0^2}}\frac{x_{r0}-x_{t0}}{c_0}\right) \Rightarrow \frac{d\tau}{dw} = -\sqrt{\frac{1+v_t/c_0}{1-v_t/c_0}}$$

$$g(t) = -\sqrt{\frac{1+v_t/c_0}{1-v_t/c_0}}\int_{\infty}^{-\infty} s\left(\sqrt{\frac{1+v_t/c_0}{1-v_t/c_0}}\left(t - w - \frac{1}{\sqrt{1-v_t^2/c_0^2}}\frac{x_{r0}-x_{t0}}{c_0}\right)\right)\delta(w)dw$$

$$g(t) = \sqrt{\frac{1+v_t/c_0}{1-v_t/c_0}} \cdot s\left(\sqrt{\frac{1+v_t/c_0}{1-v_t/c_0}}\left(t - \frac{1}{\sqrt{1-v_t^2/c_0^2}}\frac{x_{r0}-x_{t0}}{c_0}\right)\right) \quad (3.17)$$

Again the received signal is determined by time scaling, amplitude scaling and time delay. The time delay is directly affected by the Lorentz transform.

The combination of time and amplitude scaling was already mentioned 1905 by Einstein in his famous paper "Zur Elektrodynamik bewegter Körper" [2]. The authors carried out an extensive search in the literature and in the world wide web but could not find any other reference which mentions amplitude scaling caused by the Doppler Effect. It should be emphasized, that this amplitude scaling is <u>not</u> caused by variation of the distance between transmitter and receiver but only by the Doppler Effect.

3.2.2 Stationary Transmitter and uniform Motion of Receiver

Again eq. 3.5a is the starting point for the relativistic calculation. The velocity v_t of the transmitter is set to zero. An equivalent calculation could be performed starting from eq. 3.5b. Since no new insight could be gained from this calculation it is omitted.

The Lorentz transform according to eq. 3.15 is now applied to the transmission time τ of the signal, i. e. $\tau = \tau'\sqrt{1-v_r^2/c_0^2}$.

$$t_r(\tau') = \frac{\sqrt{1-v_r^2/c_0^2}}{1-v_r/c_0}\tau' + \frac{1}{1-v_r/c_0}\frac{x_{r0}-x_{t0}}{c_0} = \sqrt{\frac{1+v_r/c_0}{1-v_r/c_0}}\left(\tau' + \frac{1}{\sqrt{1-v_r^2/c_0^2}}\frac{x_{r0}-x_{t0}}{c_0}\right) \quad (3.18)$$

The calculation of the received signal is equivalent to the non-relativistic case.

$$g(t) = \int_{-\infty}^{\infty} s(\tau')\delta(t-t_r(\tau'))d\tau' = \int_{-\infty}^{\infty} s(\tau')\delta\left(t - \underbrace{\left(\sqrt{\frac{1+v_r/c_0}{1-v_r/c_0}}\left(\tau' + \frac{1}{\sqrt{1-v_r^2/c_0^2}}\frac{x_{r0}-x_{t0}}{c_0}\right)\right)}_{w}\right)d\tau'$$

A change of variables is followed by an application of the sifting property of the delta function.

$$\tau' = \sqrt{\frac{1-v_r/c_0}{1+v_r/c_0}}\left(t - w - \frac{1}{1-v_r/c_0}\frac{x_{r0}-x_{t0}}{c_0}\right) \Rightarrow \frac{d\tau'}{dw} = -\sqrt{\frac{1-v_r/c_0}{1+v_r/c_0}}$$

$$g(t) = -\sqrt{\frac{1-v_r/c_0}{1+v_r/c_0}}\int_{\infty}^{-\infty} s\left(\sqrt{\frac{1-v_r/c_0}{1+v_r/c_0}}\left(t - w - \frac{1}{1-v_r/c_0}\frac{x_{r0}-x_{t0}}{c_0}\right)\right)\delta(w)dw$$

$$g(t) = \sqrt{\frac{1-v_r/c_0}{1+v_r/c_0}}\cdot s\left(\sqrt{\frac{1-v_r/c_0}{1+v_r/c_0}}\left(t - \frac{1}{1-v_r/c_0}\frac{x_{r0}-x_{t0}}{c_0}\right)\right) \quad (3.19)$$

The time and amplitude scaling factors in eqs. 3.19 and 3.17 are identical as was to be expected in the relativistic case. Eq. 3.17 describes motion of the transmitter in the coordinate system of the receiver. Eq. 3.19 describes motion of the receiver in the coordinate system of the transmitter. This explains the different signs of the velocities of the transmitter resp. receiver.

At first glance the different time delays in eq. 3.17 and eq. 3.19 are surprising. Rewriting the scaling factors for the time delays in eqs. 3.17 and 3.19 induces the insight, that the different time delays also result from the fact that in one case the coordinate system of the receiver is used and in the other case the coordinate system of the transmitter.

$$\frac{1}{\sqrt{1-v_t^2/c_0^2}} = \sqrt{\frac{1-v_t/c_0}{1+v_t/c_0}}\frac{1}{1-v_t/c_0} \qquad \frac{1}{1-v_r/c_0} = \sqrt{\frac{1+v_r/c_0}{1-v_r/c_0}}\frac{1}{\sqrt{1-v_r^2/c_0^2}}$$

The so called Doppler factors [5] can be used to convert time delays from one coordinate system to another one in relative motion.

$$\sqrt{\frac{1-v_t/c_0}{1+v_t/c_0}} \quad \text{resp.} \quad \sqrt{\frac{1+v_r/c_0}{1-v_r/c_0}}$$

3.2.3 Uniform Motion of Transmitter and Receiver

In this case the Doppler Effect is determined by the relativistic subtraction v_d of velocities of the transmitter (v_t) and the receiver (v_r) [2, 4, 6].

$$v_d = \frac{v_t - v_r}{1 - v_t v_r/c_0^2} \quad (3.20)$$

There are two special cases which were already discussed in the preceding sections.

$$v_r = 0 \Rightarrow v_d = v_t \quad v_t = 0 \Rightarrow v_d = -v_r$$

Fig. 7 gives a graphical illustration of the range of relativistic subtractions of velocities which can result from variation of v_t resp. v_r between $-c_0$ and c_0. The relativistic subtractions of velocities according to eq. 3.20 are displayed normalised to the velocity of light c_0.

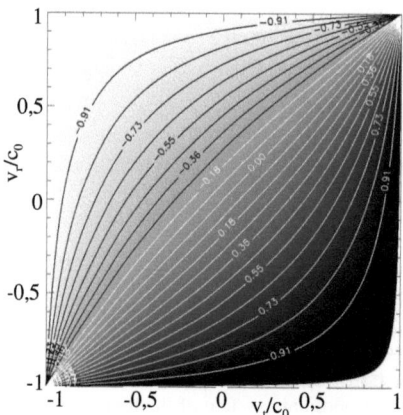

Fig. 7: Relativistic subtractions of velocities

Of course v_d is restricted to the range $-c_0 \ldots c_0$. This can be deduced from special cases.

$$v_t = c_0 \Rightarrow v_d = \frac{c_0 - v_r}{1 - c_0 v_r / c_0^2} = \frac{c_0 - v_r}{1 - v_r/c_0} = c_0 \frac{1 - v_r/c_0}{1 - v_r/c_0} = c_0$$

$$v_t = -c_0 \Rightarrow v_d = \frac{-c_0 - v_r}{1 + c_0 v_r / c_0^2} = -\frac{c_0 + v_r}{1 + v_r/c_0} = -c_0 \frac{1 + v_r/c_0}{1 + v_r/c_0} = -c_0$$

$$v_r = c_0 \Rightarrow v_d = \frac{v_t - c_0}{1 - v_t c_0 / c_0^2} = \frac{v_t - c_0}{1 - v_t/c_0} = c_0 \frac{v_t/c_0 - 1}{1 - v_t/c_0} = -c_0$$

$$v_r = -c_0 \Rightarrow v_d = \frac{v_t + c_0}{1 + v_t c_0 / c_0^2} = \frac{v_t + c_0}{1 + v_t/c_0} = c_0 \frac{1 + v_t/c_0}{1 + v_t/c_0} = c_0$$

The received signal is calculated according to eq. 3.17. The velocity v_r of the receiver is replaced by the relativistic subtraction of velocities v_d according to eq. 3.20.

$$g(t) = \sqrt{\frac{1 + v_d/c_0}{1 - v_d/c_0}} \cdot s\left(\sqrt{\frac{1 + v_d/c_0}{1 - v_d/c_0}} \left(t - \frac{1}{\sqrt{1 - v_d^2/c_0^2}} \frac{x_{r0} - x_{t0}}{c_0} \right) \right) \quad (3.21)$$

Now the received signal can be expressed as a function of v_t and v_r. First the relativistic subtraction of velocities v_d according to eq. 3.20 is normalised to the velocity of light c_0.

$$\frac{v_d}{c_0} = \frac{v_t - v_r}{c_0 - v_t v_r / c_0} \Rightarrow 1 \pm \frac{v_d}{c_0} = \frac{c_0 - v_t v_r / c_0 \pm (v_t - v_r)}{c_0 - v_t v_r / c_0}$$

The time and amplitude scaling factor according to eq. 3.21 is expressed as a function of v_t and v_r.

$$\frac{1 + v_d / c_0}{1 - v_d / c_0} = \frac{\dfrac{c_0 - v_t v_r / c_0 + (v_t - v_r)}{c_0 - v_t v_r / c_0}}{\dfrac{c_0 - v_t v_r / c_0 - (v_t - v_r)}{c_0 - v_t v_r / c_0}} \Rightarrow \sqrt{\frac{1 + v_d / c_0}{1 - v_d / c_0}} = \sqrt{\frac{c_0 - v_t v_r / c_0 + (v_t - v_r)}{c_0 - v_t v_r / c_0 - (v_t - v_r)}}$$

$$\sqrt{\frac{1 + v_d / c_0}{1 - v_d / c_0}} = \sqrt{\frac{1 - v_t v_r / c_0^2 + v_t / c_0 - v_r / c_0}{1 - v_t v_r / c_0^2 - v_t / c_0 + v_r / c_0}} = \sqrt{\frac{(1 + v_t / c_0)(1 - v_r / c_0)}{(1 + v_r / c_0)(1 - v_t / c_0)}}$$

The scaling factor for the constant time delay is also expressed as a function of v_t and v_r.

$$1 - v_d^2 / c_0^2 = (1 + v_d / c_0)(1 - v_d / c_0) = \frac{c_0 - v_t v_r / c_0 + (v_t - v_r)}{c_0 - v_t v_r / c_0} \cdot \frac{c_0 - v_t v_r / c_0 - (v_t - v_r)}{c_0 - v_t v_r / c_0}$$

$$1 - v_d^2 / c_0^2 = \frac{(c_0 - v_t v_r / c_0)^2 - (v_t - v_r)^2}{(c_0 - v_t v_r / c_0)^2} = \frac{c_0^2 - \cancel{2 v_t v_r} + (v_t v_r / c_0)^2 - v_t^2 + \cancel{2 v_t v_r} - v_r^2}{(c_0 - v_t v_r / c_0)^2}$$

$$1 - v_d^2 / c_0^2 = \frac{1 + (v_t / c_0 \, v_r / c_0)^2 - v_t^2 / c_0^2 - v_r^2 / c_0^2}{(1 - v_t / c_0 \, v_r / c_0)^2} = \frac{(1 - v_t^2 / c_0^2)(1 - v_r^2 / c_0^2)}{(1 - v_t v_r / c_0^2)^2}$$

$$\frac{1}{\sqrt{1 - v_d^2 / c_0^2}} = \frac{1 - v_t v_r / c_0^2}{\sqrt{(1 - v_t^2 / c_0^2)(1 - v_r^2 / c_0^2)}}$$

Combination of the partial results leads to a somewhat lengthy equation for the received signal.

$$g(t) = \sqrt{\frac{(1 + v_t / c_0)(1 - v_r / c_0)}{(1 + v_r / c_0)(1 - v_t / c_0)}} \cdot s\left(\sqrt{\frac{(1 + v_t / c_0)(1 - v_r / c_0)}{(1 + v_r / c_0)(1 - v_t / c_0)}} \left(t - \frac{1 - v_t v_r / c_0^2}{\sqrt{(1 - v_t^2 / c_0^2)(1 - v_r^2 / c_0^2)}} \frac{x_{r0} - x_{t0}}{c_0} \right) \right)$$

(3.22)

The plausibility of this equation can be checked considering the following arguments.

1. $g(t)$ must be a function of the relativistic subtraction of velocities only. v_t resp. v_r must not occur as separate variables. This condition is met by the approach according to eq. 3.21.

2. $g(t)$ must be a function of the difference $x_{r0} - x_{t0}$ only. x_{r0} resp. x_{t0} must not occur as separate variables. Obviously eq. 3.22 meets this condition.

3. If one of the velocities v_t resp. v_r is equal to zero eq. 3.22 must be reduced to one of the special cases according to eq. 3.17 resp. eq. 3.19.

$$v_r = 0 \Rightarrow g(t) = \sqrt{\frac{1+v_t/c_0}{1-v_t/c_0}} \cdot s\left(\sqrt{\frac{1+v_t/c_0}{1-v_t/c_0}}\left(t - \frac{1}{\sqrt{1-v_t^2/c_0^2}}\frac{x_{r0}-x_{t0}}{c_0}\right)\right)$$

This equation is identical to eq. 3.17.

$$v_t = 0 \Rightarrow g(t) = \sqrt{\frac{1-v_r/c_0}{1+v_r/c_0}} \cdot s\left(\sqrt{\frac{1-v_r/c_0}{1+v_r/c_0}}\left(t - \frac{1}{\sqrt{1-v_r^2/c_0^2}}\frac{x_{r0}-x_{t0}}{c_0}\right)\right)$$

The scaling factor for time and amplitude is the same as in eq. 3.19. The different time delay, compared to eq. 3.19, results from the fact that eq. 3.19 is correct if the coordinate system of the transmitter is used while eq. 3.22 is correct if the coordinate system of the receiver is used. Using the above mentioned Doppler factors eq. 3.22 with $v_t = 0$ can be transferred to eq. 3.19.

The converse problem arises, if the calculations are carried out using the relativistic subtraction of velocities for a stationary transmitter. Then the time delay in eq. 3.22 with $v_r = 0$ is different to the time delay in eq. 3.17. Using the above mentioned Doppler factors eq. 3.22 with $v_r = 0$ can be transferred to eq. 3.17.

4. If v_t and v_r are much smaller than the velocity of light eq. 3.22 should pass over to the non-relativistic equation 3.8a.

Using $(1+v_r/c_0)^{-1/2} \approx 1-v_r/2c_0$, $(1+v_t/c_0)^{1/2} \approx (1-v_t/2c_0)^{-1}$, $(1-v_t/c_0)^{-1/2} \approx (1+v_t/2c_0)^{-1}$ and $(1-v_r/c_0)^{1/2} \approx 1-v_r/2c_0$ and dropping second order terms leads to:

$$\sqrt{\frac{(1+v_t/c_0)(1-v_r/c_0)}{(1+v_r/c_0)(1-v_t/c_0)}} \approx \frac{1-v_r/2c_0}{1-v_t/2c_0}\frac{1-v_r/2c_0}{1-v_t/2c_0} \approx \frac{1-v_r/c_0}{1-v_t/c_0}$$

For low velocities the scaling factor for time and amplitude according to eq. 3.22 is identical to the scaling factor in eq. 3.8a.

A comparison of the time delay in eq. 3.8a (non-relativistic) and eq. 3.22 (relativistic) for low velocities is not meaningful. The time delay according to eq. 3.8a is a function of v_r only while the time delay according to eq. 3.22 also a function of v_t. The relativistic Doppler Effect is a function of the relativistic subtraction of velocities of the transmitter and the receiver. This symmetry does not occur in the non-relativistic case.

4 Uniform Motion of Transmitter and Receiver in Space

Fig. 8 schematically shows a transmitter and a receiver moving on arbitrary lines in space with constant velocities.

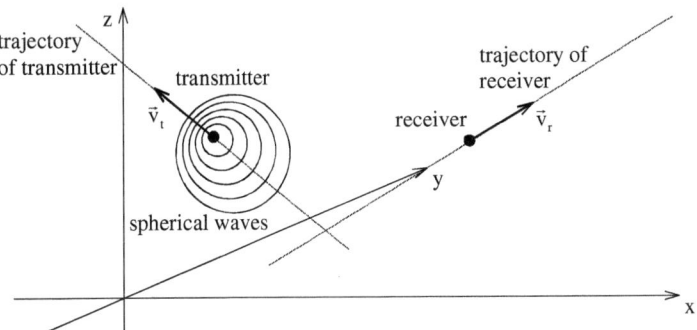

Fig. 8: Emission of signals by a transmitter moving in space

The trajectories of the transmitter and the receiver need not intersect anywhere in space.

4.1 Non-relativistic Doppler Effect

Using vector notation the locations of the transmitter resp. the receiver are functions of time according to:

$$\vec{r} = \vec{r}_{t0} + \vec{v}_t t \quad \text{resp.} \quad \vec{r} = \vec{r}_{r0} + \vec{v}_r t \tag{4.1}$$

$\vec{r}_{t0}$ resp. $\vec{r}_{r0}$ are the locations of the transmitter resp. the receiver at time $t = 0$.

At time τ the transmitter emits a spherical wave. Eq. 4.2 characterises the evolution of the sphere as a function of time. • denotes the scalar product.

$$\left(\vec{r} - (\vec{r}_{t0} + \vec{v}_t \tau)\right) \bullet \left(\vec{r} - (\vec{r}_{t0} + \vec{v}_t \tau)\right) = c_0^2 (t - \tau)^2 \tag{4.2}$$

The squared radius of the sphere at time t is on the right-hand side of the equation. The radius is proportional to the velocity of propagation c_0 of the signal.

Fig. 9 illustrates the situation.

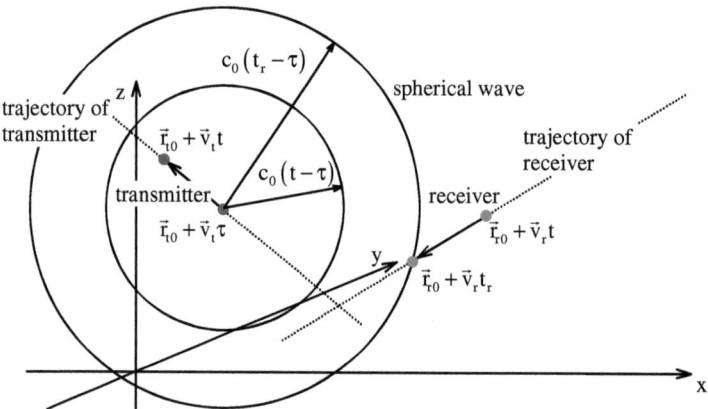

Fig. 9: Propagation of a spherical wave from the transmitter to the receiver

4.1.1   Calculation of the received Signal in General Form

At time t_r the spherical wave arrives at the receiver. This time is calculated using the trajectory of the receiver according to eq. 4.1 in eq. 4.2.

$$\left(\vec{r}_{r0}+\vec{v}_r t_r -(\vec{r}_{t0}+\vec{v}_t\tau)\right)\cdot\left(\vec{r}_{r0}+\vec{v}_r t_r -(\vec{r}_{t0}+\vec{v}_t\tau)\right) = c_0^2(t_r-\tau)^2 \quad (4.3)$$

After expansion of the scalar product and restating of eq. 4.3 the receiving time t_r can be calculated by solving the following quadratic equation.

$$t_r^2 - 2\frac{c_0^2\tau+(\vec{r}_{r0}-\vec{r}_{t0}-\vec{v}_t\tau)\cdot\vec{v}_r}{c_0^2-\vec{v}_r\cdot\vec{v}_r}t_r + \frac{c_0^2\tau^2-(\vec{r}_{r0}-\vec{r}_{t0}-\vec{v}_t\tau)\cdot(\vec{r}_{r0}-\vec{r}_{t0}-\vec{v}_t\tau)}{c_0^2-\vec{v}_r\cdot\vec{v}_r} = 0$$

$$t_r = \frac{c_0^2\tau+(\vec{r}_{r0}-\vec{r}_{t0}-\vec{v}_t\tau)\cdot\vec{v}_r}{c_0^2-\vec{v}_r\cdot\vec{v}_r} + \sqrt{\left(\frac{c_0^2\tau+(\vec{r}_{r0}-\vec{r}_{t0}-\vec{v}_t\tau)\cdot\vec{v}_r}{c_0^2-\vec{v}_r\cdot\vec{v}_r}\right)^2 - \frac{c_0^2\tau^2-(\vec{r}_{r0}-\vec{r}_{t0}-\vec{v}_t\tau)\cdot(\vec{r}_{r0}-\vec{r}_{t0}-\vec{v}_t\tau)}{c_0^2-\vec{v}_r\cdot\vec{v}_r}}$$

(4.4)

The plus sign in front of the square root is selected because of the causality condition $t_r \geq \tau$. Using eq. 4.4 the time variant impulse response $h(t,\tau)$, according to eq. 2.2, is obtained.

$$h(t,\tau) = \delta(t-t_r(\tau)) = \delta\left(t - \frac{c_0^2\tau+(\vec{r}_{r0}-\vec{r}_{t0}-\vec{v}_t\tau)\cdot\vec{v}_r}{c_0^2-\vec{v}_r\cdot\vec{v}_r}\right.$$

$$\left. - \sqrt{\left(\frac{c_0^2\tau+(\vec{r}_{r0}-\vec{r}_{t0}-\vec{v}_t\tau)\cdot\vec{v}_r}{c_0^2-\vec{v}_r\cdot\vec{v}_r}\right)^2 - \frac{c_0^2\tau^2-(\vec{r}_{r0}-\vec{r}_{t0}-\vec{v}_t\tau)\cdot(\vec{r}_{r0}-\vec{r}_{t0}-\vec{v}_t\tau)}{c_0^2-\vec{v}_r\cdot\vec{v}_r}}\right)$$

(4.5)

Basically the received signal g(t) is calculated according to the scheme used in eqs. 3.5 to 3.8: change of variables, solving for τ and subsequent application of the sifting property of the delta function. However the resulting equations are quite extensive. A simpler solution is obtained if eq. 4.3 is solved for τ directly. Subsequent application of the scheme used in eq. 3.8a leads to the final solution for the received signal g(t).

$$g(t) = \int_{-\infty}^{\infty} \left|\frac{d\tau(t-w)}{dw}\right| s(\tau(t-w))\delta(w)dw = \left|\frac{d\tau(t-w)}{dw}\right|_{w=0} s(\tau(t-w))_{w=0} = \left|-\frac{d\tau(t)}{dt}\right| \cdot s(\tau(t))$$

(4.6a)

After replacing the receiving time t_r by t, the time of emission τ is determined by solving the following quadratic equation.

$$\tau^2 - 2\frac{c_0^2 t - (\vec{r}_{r0} - \vec{r}_{t0} + \vec{v}_r t)\cdot\vec{v}_t}{c_0^2 - \vec{v}_t\cdot\vec{v}_t}\tau + \frac{c_0^2 t^2 - (\vec{r}_{r0} - \vec{r}_{t0} + \vec{v}_r t)\cdot(\vec{r}_{r0} - \vec{r}_{t0} + \vec{v}_r t)}{c_0^2 - \vec{v}_t\cdot\vec{v}_t} = 0$$

$$\tau = \frac{c_0^2 t - (\vec{r}_{r0} - \vec{r}_{t0} + \vec{v}_r t)\cdot\vec{v}_t}{c_0^2 - \vec{v}_t\cdot\vec{v}_t} - \sqrt{\left(\frac{c_0^2 t - (\vec{r}_{r0} - \vec{r}_{t0} + \vec{v}_r t)\cdot\vec{v}_t}{c_0^2 - \vec{v}_t\cdot\vec{v}_t}\right)^2 - \frac{c_0^2 t^2 - (\vec{r}_{r0} - \vec{r}_{t0} + \vec{v}_r t)\cdot(\vec{r}_{r0} - \vec{r}_{t0} + \vec{v}_r t)}{c_0^2 - \vec{v}_t\cdot\vec{v}_t}}$$

(4.6b)

The minus sign in front of the square root is selected because of the causality condition $\tau \leq t$ (t_r). Now τ is differentiated with respect to t.

$$\frac{d\tau}{dt} = \frac{c_0^2 - \vec{v}_r\cdot\vec{v}_t}{c_0^2 - \vec{v}_t\cdot\vec{v}_t} - \frac{\dfrac{c_0^2 t - (\vec{r}_{r0} - \vec{r}_{t0} + \vec{v}_r t)\cdot\vec{v}_t}{c_0^2 - \vec{v}_t\cdot\vec{v}_t} \cdot \dfrac{c_0^2 - \vec{v}_r\cdot\vec{v}_t}{c_0^2 - \vec{v}_t\cdot\vec{v}_t} - \dfrac{c_0^2 t - (\vec{r}_{r0} - \vec{r}_{t0} + \vec{v}_r t)\cdot\vec{v}_r}{c_0^2 - \vec{v}_t\cdot\vec{v}_t}}{\sqrt{\left(\dfrac{c_0^2 t - (\vec{r}_{r0} - \vec{r}_{t0} + \vec{v}_r t)\cdot\vec{v}_t}{c_0^2 - \vec{v}_t\cdot\vec{v}_t}\right)^2 - \dfrac{c_0^2 t^2 - (\vec{r}_{r0} - \vec{r}_{t0} + \vec{v}_r t)\cdot(\vec{r}_{r0} - \vec{r}_{t0} + \vec{v}_r t)}{c_0^2 - \vec{v}_t\cdot\vec{v}_t}}}$$

(4.6c)

Using eqs. 4.6b and 4.6c in eq. 4.6a a closed-form solution for the received signal is obtained. However this solution appears to be very large and not very clear.

$$g(t) = \left|\frac{c_0^2 - \vec{v}_r\cdot\vec{v}_t}{c_0^2 - \vec{v}_t\cdot\vec{v}_t} - \frac{\dfrac{c_0^2 t - (\vec{r}_{r0} - \vec{r}_{t0} + \vec{v}_r t)\cdot\vec{v}_t}{c_0^2 - \vec{v}_t\cdot\vec{v}_t} \cdot \dfrac{c_0^2 - \vec{v}_r\cdot\vec{v}_t}{c_0^2 - \vec{v}_t\cdot\vec{v}_t} - \dfrac{c_0^2 t - (\vec{r}_{r0} - \vec{r}_{t0} + \vec{v}_r t)\cdot\vec{v}_r}{c_0^2 - \vec{v}_t\cdot\vec{v}_t}}{\sqrt{\left(\dfrac{c_0^2 t - (\vec{r}_{r0} - \vec{r}_{t0} + \vec{v}_r t)\cdot\vec{v}_t}{c_0^2 - \vec{v}_t\cdot\vec{v}_t}\right)^2 - \dfrac{c_0^2 t^2 - (\vec{r}_{r0} - \vec{r}_{t0} + \vec{v}_r t)\cdot(\vec{r}_{r0} - \vec{r}_{t0} + \vec{v}_r t)}{c_0^2 - \vec{v}_t\cdot\vec{v}_t}}}\right|$$

$$\cdot s\left(\frac{c_0^2 t - (\vec{r}_{r0} - \vec{r}_{t0} + \vec{v}_r t)\cdot\vec{v}_t}{c_0^2 - \vec{v}_t\cdot\vec{v}_t} - \sqrt{\left(\frac{c_0^2 t - (\vec{r}_{r0} - \vec{r}_{t0} + \vec{v}_r t)\cdot\vec{v}_t}{c_0^2 - \vec{v}_t\cdot\vec{v}_t}\right)^2 - \frac{c_0^2 t^2 - (\vec{r}_{r0} - \vec{r}_{t0} + \vec{v}_r t)\cdot(\vec{r}_{r0} - \vec{r}_{t0} + \vec{v}_r t)}{c_0^2 - \vec{v}_t\cdot\vec{v}_t}}\right)$$

Further insights can be gained from special cases.

4.1.2 Stationary Receiver

First a stationary receiver located at the point of origin is considered, i.e. $\vec{v}_r = \vec{0}$, $\vec{r}_{r0} = \vec{0}$. The transmitter moves with constant velocity on a line parallel to the x-axis of the coordinate system, i.e. $\vec{r}_{t0} = (0, y_0, 0)$, $\vec{v}_t = (v_t, 0, 0)$. Without loss of generality all z-components are set to zero. Fig. 10 illustrates this configuration. This special case can be used to investigate the transverse Doppler Effect (motion of transmitter transverse to the receiver) and the longitudinal Doppler Effect (motion of transmitter in the direction of the receiver).

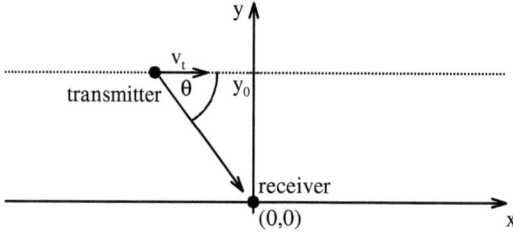

Fig. 10: Transmitter passing a receiver located at the point of origin

Using the above mentioned simplifications eqs. 4.6b resp. 4.6c are reduced to:

$$\tau = \frac{c_0^2 t}{c_0^2 - v_t^2} - \sqrt{\left(\frac{c_0^2 t}{c_0^2 - v_t^2}\right)^2 - \frac{c_0^2 t^2 - y_0^2}{c_0^2 - v_t^2}} = \frac{c_0^2 t - \sqrt{v_t^2 c_0^2 t^2 + (c_0^2 - v_t^2) y_0^2}}{c_0^2 - v_t^2} \tag{4.7a}$$

$$\frac{d\tau}{dt} = \frac{c_0^2}{c_0^2 - v_t^2} \left(1 - \frac{v_t^2 t}{\sqrt{v_t^2 c_0^2 t^2 + (c_0^2 - v_t^2) y_0^2}}\right) \tag{4.7b}$$

In this special case, the received signal g(t) is described by a much simpler equation.

$$g(t) = \left|\frac{c_0^2}{c_0^2 - v_t^2}\left(1 - \frac{v_t^2 t}{\sqrt{v_t^2 c_0^2 t^2 + (c_0^2 - v_t^2) y_0^2}}\right)\right| \cdot s\left(\frac{c_0^2 t - \sqrt{v_t^2 c_0^2 t^2 + (c_0^2 - v_t^2) y_0^2}}{c_0^2 - v_t^2}\right) \tag{4.7c}$$

Now the transmitted signal is a complex sinusoidal waveform $s(t) = \exp(j2\pi f_0 t)$. Eq. 4.7c enables the calculation of the instantaneous frequency of the received signal.

$$g(t) = \left|\frac{c_0^2}{c_0^2 - v_t^2}\left(1 - \frac{v_t^2 t}{\sqrt{v_t^2 c_0^2 t^2 + (c_0^2 - v_t^2) y_0^2}}\right)\right| \cdot \exp\left(j2\pi f_0 \underbrace{\frac{c_0^2 t - \sqrt{v_t^2 c_0^2 t^2 + (c_0^2 - v_t^2) y_0^2}}{c_0^2 - v_t^2}}_{\varphi(t)}\right)$$

The instantaneous frequency f(t) of the received signal is directly proportional to the derivative of its phase φ(t) with respect to time, i. e.:

$$f(t) = \frac{1}{2\pi} \frac{d\varphi(t)}{dt} = f_0 \frac{c_0^2}{c_0^2 - v_t^2} \left(1 - \frac{v_t^2 t}{\sqrt{v_t^2 c_0^2 t^2 + (c_0^2 - v_t^2) y_0^2}} \right) \quad (4.8)$$

From two limiting cases frequency scaling factors, according to eq. 3.10, are obtained.

$$t \to \infty \Rightarrow f(\infty) = f_0 \frac{1}{1 + v_t/c_0} \qquad t \to -\infty \Rightarrow f(-\infty) = f_0 \frac{1}{1 - v_t/c_0}$$

Fig. 11 shows some examples of normalised instantaneous frequencies as functions of time. The parameters $y_0 = 100$ km and $c_0 = 300.000$ km/s were used for the numerical calculations.

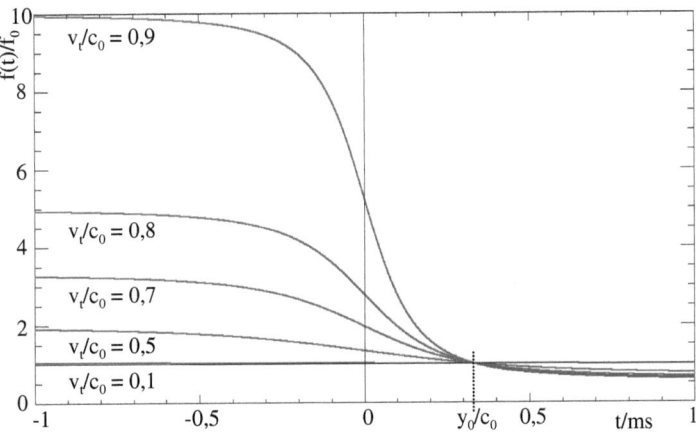

Fig. 11: Instantaneous frequencies of received signals while the transmitter passes a receiver located at the point of origin, non-relativistic case

All curves intersect at time t_0. At this time all instantaneous frequencies are equal to the frequency f_0 of the transmitted signal. t_0 is calculated using eq. 4.8.

$$f_0 = f_0 \frac{c_0^2}{c_0^2 - v_t^2} \left(1 - \frac{v_t^2 t_0}{\sqrt{v_t^2 c_0^2 t_0^2 + (c_0^2 - v_t^2) y_0^2}} \right) \Rightarrow c_0^2 \frac{v_t^2 t_0}{\sqrt{v_t^2 c_0^2 t_0^2 + (c_0^2 - v_t^2) y_0^2}} = v_t^2$$

$$c_0^2 v_t^2 t_0 = v_t^2 \sqrt{v_t^2 c_0^2 t_0^2 + (c_0^2 - v_t^2) y_0^2} \Rightarrow c_0^4 t_0^2 = v_t^2 c_0^2 t_0^2 + (c_0^2 - v_t^2) y_0^2$$

$$t_0 = y_0 / c_0 \quad (4.9)$$

The fraction of the signal which arrives at the receiver at time t_0 was transmitted at time $t = 0$ when the x-coordinate of the transmitter was equal to 0. At $t = 0$ the motion vector of the transmitter is perpendicular to the line of sight from the transmitter to the receiver. It is well known from literature on the Doppler Effect that the non-relativistic Doppler Effect only depends on the longitudinal component of the motion vector of the transmitter [4].

4.1.3 Stationary Transmitter

Now a stationary transmitter located at the point of origin is considered, i. e. $\vec{v}_t = \vec{0}$, $\vec{r}_{t0} = \vec{0}$. The receiver moves with constant velocity on a line parallel to the x-axis of the coordinate system, i. e. $\vec{r}_{r0} = (0, y_0, 0)$, $\vec{v}_r = (v_r, 0, 0)$. Without loss of generality all z-components are set to zero. Fig. 12 illustrates this configuration.

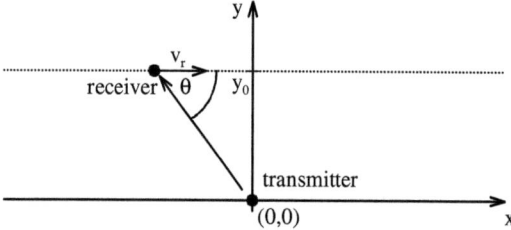

Fig. 12: Receiver passing a transmitter located at the point of origin

Using the above mentioned simplifications eqs. 4.6b resp. 4.6c are reduced to:

$$\tau = t - c_0^{-1}\sqrt{v_r^2 t^2 + y_0^2} \tag{4.10a}$$

$$\frac{d\tau}{dt} = 1 - c_0^{-1}\frac{v_r^2 t}{\sqrt{v_r^2 t^2 + y_0^2}} \tag{4.10b}$$

$$\Rightarrow g(t) = \left|1 - c_0^{-1}\frac{v_r^2 t}{\sqrt{v_r^2 t^2 + y_0^2}}\right| \cdot s\left(t - c_0^{-1}\sqrt{v_r^2 t^2 + y_0^2}\right) \tag{4.10c}$$

Again the transmitted signal is a complex sinusoidal waveform $s(t) = \exp(j2\pi f_0 t)$.

$$g(t) = \left|1 - c_0^{-1}\frac{v_r^2 t}{\sqrt{v_r^2 t^2 + y_0^2}}\right| \cdot \exp\underbrace{\left(j2\pi f_0 \left(t - c_0^{-1}\sqrt{v_r^2 t^2 + y_0^2}\right)\right)}_{\varphi(t)}$$

The instantaneous frequency f(t) of the received signal is calculated using the derivative of its phase φ(t) with respect to time.

$$f(t) = \frac{1}{2\pi}\frac{d}{dt}2\pi f_0\left(t - c_0^{-1}\sqrt{v_r^2 t^2 + y_0^2}\right) = f_0\left(1 - c_0^{-1}\frac{v_r^2 t}{\sqrt{v_r^2 t^2 + y_0^2}}\right) \quad (4.11)$$

From two limiting cases frequency scaling factors, according to eq. 3.9, are obtained.

$$t \to \infty \Rightarrow f(\infty) = f_0\left(1 - \frac{v_r}{c_0}\right) \qquad t \to -\infty \Rightarrow f(-\infty) = f_0\left(1 + \frac{v_r}{c_0}\right)$$

Fig. 13 shows some examples of normalised instantaneous frequencies as functions of time. The parameters $y_0 = 100$ km and $c_0 = 300.000$ km/s were used for the numerical calculations.

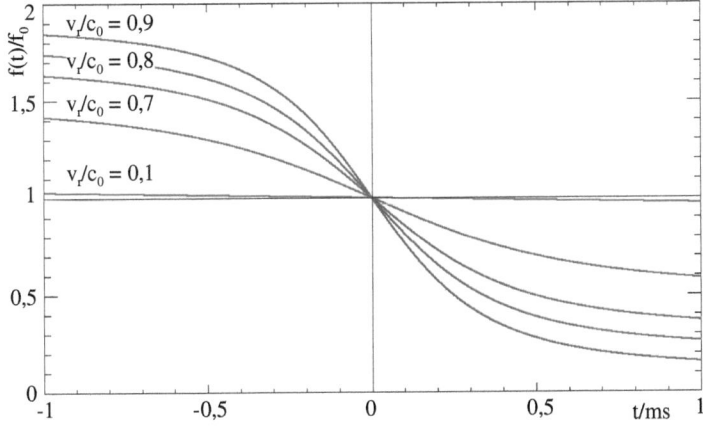

Fig. 13: Instantaneous frequencies of received signals while the receiver passes a transmitter located at the point of origin, non-relativistic case

The curves in fig. 13 differ considerably from the curves in fig. 11. In contrast to the relativistic Doppler Effect the non-relativistic Doppler Effect is asymmetric with respect to motion of the transmitter or the receiver. Eqs. 1.1 and 1.2 clarify this effect.

All curves intersect at time t_0. At this time all instantaneous frequencies are equal to the frequency f_0 of the transmitted signal. $t_0 = 0$ is easily calculated using eq. 4.11.

The fraction of the signal which arrives at the receiver at time $t_0 = 0$ was transmitted at time $-y_0/c_0$ when the x-coordinate of the receiver was equal to 0. At $t = 0$ the motion vector of the receiver is perpendicular to the line of sight from the transmitter to the receiver. It is well known from literature on the Doppler Effect that the non-relativistic Doppler Effect only depends on the longitudinal component of the motion vector [4].

4.1.4 Auto Correlation Function (ACF) of the received Signal

The calculation of the ACF $\varphi_{gg}(\Delta t)$ of the received signal g(t) according to eq. 3.6 is demonstrated without loss of generality for the special case of a stationary receiver $(\vec{v}_r = \vec{0})$ located at $\vec{r}_{r0} = \vec{0}$. $\vec{r}_{t0}$ is replaced by $\vec{r}_0$ for brevity.

$$\varphi_{gg}(\Delta t) = \int_{-\infty}^{\infty} g(t)g(t-\Delta t)dt = \int_{-\infty}^{\infty}\int_{-\infty}^{\infty} s(\tau)\delta\left(t-\left(\tau+c_0^{-1}\sqrt{(\vec{r}_0+\vec{v}_t\tau)\bullet(\vec{r}_0+\vec{v}_t\tau)}\right)\right)d\tau$$

$$\cdot \int_{-\infty}^{\infty} s(\hat{\tau})\delta\underbrace{\left(t-\Delta t-\left(\hat{\tau}+c_0^{-1}\sqrt{(\vec{r}_0+\vec{v}_t\hat{\tau})\bullet(\vec{r}_0+\vec{v}_t\hat{\tau})}\right)\right)}_{u}d\hat{\tau}dt \tag{4.12}$$

Using two subsequent changes of variables the ACF of the received signal can be expressed as a function of the transmitted signal s(t).

First change of variables: $\quad t = u+\Delta t+\hat{\tau}+c_0^{-1}\sqrt{(\vec{r}_0+\vec{v}_t\hat{\tau})\bullet(\vec{r}_0+\vec{v}_t\hat{\tau})} \Rightarrow dt = du$

$$\varphi_{gg}(\Delta t) = \int_{-\infty}^{\infty}\int_{-\infty}^{\infty}\int_{-\infty}^{\infty} s(\tau)s(\hat{\tau})\delta(u)$$

$$\cdot\delta\left(u+\Delta t+\hat{\tau}-\tau+c_0^{-1}\sqrt{(\vec{r}_0+\vec{v}_t\hat{\tau})\bullet(\vec{r}_0+\vec{v}_t\hat{\tau})}-c_0^{-1}\sqrt{(\vec{r}_0+\vec{v}_t\tau)\bullet(\vec{r}_0+\vec{v}_t\tau)}\right)d\tau d\hat{\tau} du$$

$$\varphi_{gg}(\Delta t) = \int_{-\infty}^{\infty}\int_{-\infty}^{\infty} s(\tau)s(\hat{\tau})\delta\underbrace{\left(\Delta t+\hat{\tau}-\tau+c_0^{-1}\sqrt{(\vec{r}_0+\vec{v}_t\hat{\tau})\bullet(\vec{r}_0+\vec{v}_t\hat{\tau})}-c_0^{-1}\sqrt{(\vec{r}_0+\vec{v}_t\tau)\bullet(\vec{r}_0+\vec{v}_t\tau)}\right)}_{w}d\tau d\hat{\tau}$$

Second change of variables:

$$\sqrt{(\vec{r}_0+\vec{v}_t\hat{\tau})\bullet(\vec{r}_0+\vec{v}_t\hat{\tau})} = c_0(w-\Delta t+\tau)+\sqrt{(\vec{r}_0+\vec{v}_t\tau)\bullet(\vec{r}_0+\vec{v}_t\tau)}-c_0\hat{\tau}$$

$$\hat{\tau}^2 - 2\frac{c_0^2(w-\Delta t+\tau)+c_0\sqrt{(\vec{r}_0+\vec{v}_t\tau)\bullet(\vec{r}_0+\vec{v}_t\tau)}+\vec{r}_0\bullet\vec{v}_t}{c_0^2-\vec{v}_t\bullet\vec{v}_t}\hat{\tau}$$

$$+\frac{\left(c_0(w-\Delta t+\tau)+\sqrt{(\vec{r}_0+\vec{v}_t\tau)\bullet(\vec{r}_0+\vec{v}_t\tau)}\right)^2-\vec{r}_0\bullet\vec{r}_0}{c_0^2-\vec{v}_t\bullet\vec{v}_t} = 0 \tag{4.13}$$

The solution of the quadratic equation 4.13 for $\hat{\tau}$ is quite extensive. Therefore the calculation of $\varphi_{gg}(\Delta t)$ is cumbersome.

$$\hat{\tau} = \frac{c_0^2(w-\Delta t+\tau)+c_0\sqrt{(\vec{r}_0+\vec{v}_t\tau)\bullet(\vec{r}_0+\vec{v}_t\tau)}+\vec{r}_0\bullet\vec{v}_t}{c_0^2-\vec{v}_t\bullet\vec{v}_t} \pm$$

$$\sqrt{\left(\frac{c_0^2(w-\Delta t+\tau)+c_0\sqrt{(\vec{r}_0+\vec{v}_t\tau)\bullet(\vec{r}_0+\vec{v}_t\tau)}+\vec{r}_0\bullet\vec{v}_t}{c_0^2-\vec{v}_t\bullet\vec{v}_t}\right)^2 - \frac{\left(c_0(w-\Delta t+\tau)+\sqrt{(\vec{r}_0+\vec{v}_t\tau)\bullet(\vec{r}_0+\vec{v}_t\tau)}\right)^2-\vec{r}_0\bullet\vec{r}_0}{c_0^2-\vec{v}_t\bullet\vec{v}_t}}$$

(4.14a)

Even after some reduction, the equation for the amplitude scaling factor $|d\hat{\tau}/dw|$ stays quite bulky.

$$\frac{d\hat{\tau}}{dw} = \frac{c_0^2}{c_0^2-\vec{v}_t\bullet\vec{v}_t} \pm$$

$$\frac{\dfrac{c_0^2}{(c_0^2-\vec{v}_t\bullet\vec{v}_t)^2}\left(\dfrac{\vec{r}_0\bullet\vec{v}_t}{c_0^2-\vec{v}_t\bullet\vec{v}_t}+\dfrac{\vec{v}_t\bullet\vec{v}_t}{c_0^2-\vec{v}_t\bullet\vec{v}_t}\left(w-\Delta t+\tau+\dfrac{\sqrt{(\vec{r}_0+\vec{v}_t\tau)\bullet(\vec{r}_0+\vec{v}_t\tau)}}{c_0}\right)\right)}{\sqrt{\left(\dfrac{c_0^2(w-\Delta t+\tau)+c_0\sqrt{(\vec{r}_0+\vec{v}_t\tau)\bullet(\vec{r}_0+\vec{v}_t\tau)}+\vec{r}_0\bullet\vec{v}_t}{c_0^2-\vec{v}_t\bullet\vec{v}_t}\right)^2 - \dfrac{\left(c_0(w-\Delta t+\tau)+\sqrt{(\vec{r}_0+\vec{v}_t\tau)\bullet(\vec{r}_0+\vec{v}_t\tau)}\right)^2-\vec{r}_0\bullet\vec{r}_0}{c_0^2-\vec{v}_t\bullet\vec{v}_t}}}$$

(4.14b)

In general the ACF $\varphi_{gg}(\Delta t)$ of the received signal g(t) can not be expressed as a direct function of the ACF $\varphi_{ss}(\Delta t)$ of the transmitted signal s(t).

$$\varphi_{gg}(\Delta t) = \int_{-\infty}^{\infty}\int_{-\infty}^{\infty}\left|\frac{d\hat{\tau}(\tau,\Delta t,w)}{dw}\right|s(\tau)s(\hat{\tau}(\tau,\Delta t,w))\delta(w)dwd\tau = \int_{-\infty}^{\infty}\left|\frac{d\hat{\tau}(\tau,\Delta t,w)}{dw}\right|_{w=0}s(\tau)s(\hat{\tau}(\tau,\Delta t,w)_{w=0})d\tau$$

(4.14c)

The application of the derivation of the Doppler Effect in terms of system theory to stochastic signals is restricted to special cases such as uniform motion of the transmitter and the receiver on a line.

4.1.5 Fourier Transform of the received Signal

The calculation of the Fourier transform of the received signal according to eq. 4.6 is demonstrated without loss of generality for the special case of a stationary receiver located at the point of origin. Using $\vec{v}_r = \vec{0}$ and $\vec{r}_{r0} = \vec{0}$ the Fourier transform is calculated. $\vec{r}_{t0}$ is replaced by $\vec{r}_0$ for brevity.

$$\underline{G}(f) = \int_{-\infty}^{\infty} g(t)e^{-j2\pi ft} dt = \int_{-\infty}^{\infty} \left| \frac{c_0^2}{c_0^2 - \vec{v}_t \cdot \vec{v}_t} - \frac{\frac{\cancel{\partial t}}{c_0^2 - \vec{v}_t \cdot \vec{v}_t} + \frac{\vec{r}_0 \cdot \vec{v}_t}{c_0^2 - \vec{v}_t \cdot \vec{v}_t} \quad \frac{c_0^2}{c_0^2 - \vec{v}_t \cdot \vec{v}_t} \quad \frac{\cancel{\partial t} - \vec{v}_t \cdot \vec{v}_t}{c_0^2 - \vec{v}_t \cdot \vec{v}_t} \quad \frac{c_0^2 t}{c_0^2 - \vec{v}_t \cdot \vec{v}_t}}{\sqrt{\left(\frac{c_0^2 t + \vec{r}_0 \cdot \vec{v}_t}{c_0^2 - \vec{v}_t \cdot \vec{v}_t}\right)^2 - \frac{c_0^2 t^2 - \vec{r}_0 \cdot \vec{r}_0}{c_0^2 - \vec{v}_t \cdot \vec{v}_t}}} \right|$$

$$\cdot s\left(\frac{c_0^2 t + \vec{r}_0 \cdot \vec{v}_t - \sqrt{(c_0^2 t + \vec{r}_0 \cdot \vec{v}_t)^2 - (c_0^2 - \vec{v}_t \cdot \vec{v}_t)(c_0^2 t^2 - \vec{r}_0 \cdot \vec{r}_0)}}{c_0^2 - \vec{v}_t \cdot \vec{v}_t} \right) e^{-j2\pi ft} dt \quad (4.15)$$

A change of variables is applied to the argument of the transmitted signal s(...).

$$\underline{G}(f) = \int_{-\infty}^{\infty} \left| \frac{c_0^2}{c_0^2 - \vec{v}_t \cdot \vec{v}_t} \left(1 - \frac{\vec{r}_0 \cdot \vec{v}_t + \vec{v}_t \cdot \vec{v}_t \, t}{\sqrt{(c_0^2 t + \vec{r}_0 \cdot \vec{v}_t)^2 - (c_0^2 - \vec{v}_t \cdot \vec{v}_t)(c_0^2 t^2 - \vec{r}_0 \cdot \vec{r}_0)}} \right) \right|$$

$$\cdot s\underbrace{\left(\frac{c_0^2 t + \vec{r}_0 \cdot \vec{v}_t - \sqrt{(c_0^2 t + \vec{r}_0 \cdot \vec{v}_t)^2 - (c_0^2 - \vec{v}_t \cdot \vec{v}_t)(c_0^2 t^2 - \vec{r}_0 \cdot \vec{r}_0)}}{c_0^2 - \vec{v}_t \cdot \vec{v}_t} \right)}_{u} e^{-j2\pi ft} dt$$

The expression for the new variable u is solved for t.

$$t^2 - 2ut + \frac{(c_0^2 - \vec{v}_t \cdot \vec{v}_t)u^2 - 2\vec{r}_0 \cdot \vec{v}_t u - \vec{r}_0 \cdot \vec{r}_0}{c_0^2} = 0$$

$$t = u \pm c_0^{-1} \sqrt{(\vec{r}_0 + \vec{v}_t u) \cdot (\vec{r}_0 + \vec{v}_t u)} \quad (4.16a)$$

$$\frac{dt}{du} = 1 \pm c_0^{-1} \frac{\vec{v}_t \cdot (\vec{r}_0 + \vec{v}_t u)}{\sqrt{(\vec{r}_0 + \vec{v}_t u) \cdot (\vec{r}_0 + \vec{v}_t u)}} \quad (4.16b)$$

In the final equation t(u) is used as abbreviation for eq. 4.16a. Otherwise eq. 4.17 would become extremely bulky. Unfortunately, hardly any reduction is applicable.

$$\underline{G}(f) = \int_{-\infty}^{\infty} \left| \frac{c_0^2}{c_0^2 - \vec{v}_t \cdot \vec{v}_t} \left(1 - \frac{\vec{r}_0 \cdot \vec{v}_t + \vec{v}_t \cdot \vec{v}_t \, t(u)}{\sqrt{(c_0^2 t(u) + \vec{r}_0 \cdot \vec{v}_t)^2 - (c_0^2 - \vec{v}_t \cdot \vec{v}_t)(c_0^2 t^2(u) - \vec{r}_0 \cdot \vec{r}_0)}} \right) \right| \left| \frac{dt(u)}{du} \right| s(u) e^{-j2\pi ft(u)} du$$

$$\underline{G}(f) = \int_{-\infty}^{\infty} \left| \frac{c_0^2}{c_0^2 - \vec{v}_t \cdot \vec{v}_t} \left(1 - \frac{\vec{v}_t \cdot (\vec{r}_0 + \vec{v}_t t(u))}{\sqrt{(c_0^2 t(u) + \vec{r}_0 \cdot \vec{v}_t)^2 - (c_0^2 - \vec{v}_t \cdot \vec{v}_t)(c_0^2 t^2(u) - \vec{r}_0 \cdot \vec{r}_0)}} \right) \right|$$
$$\cdot \left| 1 + \frac{1}{c_0} \frac{\vec{v}_t \cdot (\vec{r}_0 + \vec{v}_t u)}{\sqrt{(\vec{r}_0 + \vec{v}_t u) \cdot (\vec{r}_0 + \vec{v}_t u)}} \right| \cdot s(u) e^{-j2\pi f \left(u + \frac{\sqrt{(\vec{r}_0 + \vec{v}_t u) \cdot (\vec{r}_0 + \vec{v}_t u)}}{c_0} \right)} du \quad (4.17)$$

In general the Fourier transform $\underline{G}(f)$ of the received signal can not be expressed as a direct function of the Fourier transform $\underline{S}(f)$ of the transmitted signal. The transmitted signal is multiplied by a complicated function of time and there is also a non-linear time scaling in the exponent. It is extremely unlikely to find a closed form solution for the integral in eq. 4.17 for a particular transmitted signal s(t).

4.2 Relativistic Doppler Effect

The relativistic Doppler Effect for arbitrary uniform motion of the transmitter and/or the receiver is derived using the simplified purely temporal Lorentz transform [4] as in section 3.2.

4.2.1 Stationary Receiver

First the special case of a stationary receiver and uniform motion of the transmitter is investigated. Eq. 4.4 is the starting point for the relativistic calculations. The velocity $\vec{v}_r$ of the receiver is set to zero. The equation for the calculation of the receiving time t_r is now much simpler.

$$t_r = \tau + c_0^{-1} \sqrt{(\vec{v}_t \tau + \vec{r}_{t0} - \vec{r}_{r0}) \cdot (\vec{v}_t \tau + \vec{r}_{t0} - \vec{r}_{r0})} \quad (4.18)$$

The length of the subsequent equations is further reduced using $\vec{r}_{t0} - \vec{r}_{r0} = \vec{r}_0$. Then the Lorentz transform according to eq. 3.15 is applied to the receiving time t_r.

$$t_r' = \frac{t_r(\tau)}{\sqrt{1 - \vec{v}_t \cdot \vec{v}_t / c_0^2}} = \frac{c_0 \tau + \sqrt{(\vec{v}_t \tau + \vec{r}_0) \cdot (\vec{v}_t \tau + \vec{r}_0)}}{\sqrt{c_0^2 - \vec{v}_t \cdot \vec{v}_t}} \quad (4.19)$$

The linear time variant impulse response $h(t,\tau)$, according to eq. 2.2, is obtained from this equation.

$$h(t,\tau) = \delta(t - t_r'(\tau)) = \delta\left(t - \frac{c_0 \tau + \sqrt{(\vec{v}_t \tau + \vec{r}_0) \cdot (\vec{v}_t \tau + \vec{r}_0)}}{\sqrt{c_0^2 - \vec{v}_t \cdot \vec{v}_t}} \right)$$

The scheme used in eq. 4.6 enables a more efficient calculation of the received signal. After replacing of t'_r by t eq. 4.19 is directly solved for τ.

$$(\vec{v}_t\tau+\vec{r}_0)\cdot(\vec{v}_t\tau+\vec{r}_0) = \left(t\sqrt{c_0^2-\vec{v}_t\cdot\vec{v}_t} - c_0\tau\right)^2$$

$$\vec{v}_t\cdot\vec{v}_t\tau^2 + 2\vec{r}_0\cdot\vec{v}_t\tau + \vec{r}_0\cdot\vec{r}_0 = c_0^2\tau^2 - 2c_0 t\sqrt{c_0^2-\vec{v}_t\cdot\vec{v}_t}\,\tau + \left(c_0^2-\vec{v}_t\cdot\vec{v}_t\right)t^2$$

$$\tau^2 - 2\frac{c_0 t\sqrt{c_0^2-\vec{v}_t\cdot\vec{v}_t}+\vec{r}_0\cdot\vec{v}_t}{c_0^2-\vec{v}_t\cdot\vec{v}_t}\tau + \frac{\left(c_0^2-\vec{v}_t\cdot\vec{v}_t\right)t^2-\vec{r}_0\cdot\vec{r}_0}{c_0^2-\vec{v}_t\cdot\vec{v}_t} = 0$$

This quadratic equation is solved for τ.

$$\tau = \frac{\sqrt{c_0^2-\vec{v}_t\cdot\vec{v}_t}\,c_0 t + \vec{r}_0\cdot\vec{v}_t}{c_0^2-\vec{v}_t\cdot\vec{v}_t} - \sqrt{\left(\frac{\sqrt{c_0^2-\vec{v}_t\cdot\vec{v}_t}\,c_0 t + \vec{r}_0\cdot\vec{v}_t}{c_0^2-\vec{v}_t\cdot\vec{v}_t}\right)^2 - \frac{\left(c_0^2-\vec{v}_t\cdot\vec{v}_t\right)t^2-\vec{r}_0\cdot\vec{r}_0}{c_0^2-\vec{v}_t\cdot\vec{v}_t}} \qquad (4.20a)$$

The minus sign in front of the square root is selected because of the causality condition $\tau \leq t$. Now τ is differentiated with respect to t.

$$\frac{d\tau}{dt} = \frac{c_0}{\sqrt{c_0^2-\vec{v}_t\cdot\vec{v}_t}} + \frac{t - \dfrac{\sqrt{c_0^2-\vec{v}_t\cdot\vec{v}_t}\,c_0 t + \vec{r}_0\cdot\vec{v}_t}{c_0^2-\vec{v}_t\cdot\vec{v}_t}\cdot\dfrac{c_0}{\sqrt{c_0^2-\vec{v}_t\cdot\vec{v}_t}}}{\sqrt{\left(\dfrac{\sqrt{c_0^2-\vec{v}_t\cdot\vec{v}_t}\,c_0 t + \vec{r}_0\cdot\vec{v}_t}{c_0^2-\vec{v}_t\cdot\vec{v}_t}\right)^2 - \dfrac{\left(c_0^2-\vec{v}_t\cdot\vec{v}_t\right)t^2-\vec{r}_0\cdot\vec{r}_0}{c_0^2-\vec{v}_t\cdot\vec{v}_t}}} \qquad (4.20b)$$

The received signal g(t) is obtained by combination of eqs. 4.20a and 4.20b.

$$g(t) = \left| \frac{c_0}{\sqrt{c_0^2-\vec{v}_t\cdot\vec{v}_t}} + \frac{t - \dfrac{\sqrt{c_0^2-\vec{v}_t\cdot\vec{v}_t}\,c_0 t + \vec{r}_0\cdot\vec{v}_t}{c_0^2-\vec{v}_t\cdot\vec{v}_t}\cdot\dfrac{c_0}{\sqrt{c_0^2-\vec{v}_t\cdot\vec{v}_t}}}{\sqrt{\left(\dfrac{\sqrt{c_0^2-\vec{v}_t\cdot\vec{v}_t}\,c_0 t + \vec{r}_0\cdot\vec{v}_t}{c_0^2-\vec{v}_t\cdot\vec{v}_t}\right)^2 - \dfrac{\left(c_0^2-\vec{v}_t\cdot\vec{v}_t\right)t^2-\vec{r}_0\cdot\vec{r}_0}{c_0^2-\vec{v}_t\cdot\vec{v}_t}}} \right| \qquad (4.20c)$$

$$\cdot s\left(\frac{\sqrt{c_0^2-\vec{v}_t\cdot\vec{v}_t}\,c_0 t + \vec{r}_0\cdot\vec{v}_t}{c_0^2-\vec{v}_t\cdot\vec{v}_t} - \sqrt{\left(\frac{\sqrt{c_0^2-\vec{v}_t\cdot\vec{v}_t}\,c_0 t + \vec{r}_0\cdot\vec{v}_t}{c_0^2-\vec{v}_t\cdot\vec{v}_t}\right)^2 - \frac{\left(c_0^2-\vec{v}_t\cdot\vec{v}_t\right)t^2-\vec{r}_0\cdot\vec{r}_0}{c_0^2-\vec{v}_t\cdot\vec{v}_t}}\right)$$

Further insights can be gained from special cases. Without loss of generality the same simplifications as in the non-relativistic case are used. The transmitter is moving on a line parallel to the x-axis of the coordinate system, i. e. $\vec{r}_0 = (0, y_0, 0)$ resp. $\vec{v}_t = (v_t, 0, 0)$. Such eq. 4.20 can be replaced by a much simpler equation. Fig. 10 illustrates this configuration.

$$g(t) = \frac{c_0}{\sqrt{c_0^2 - v_t^2}} \left| 1 - \frac{v_t}{c_0} \frac{v_t t}{\sqrt{v_t^2 t^2 + y_0^2}} \right| \cdot s\left(\frac{c_0}{\sqrt{c_0^2 - v_t^2}} \left(t - c_0^{-1}\sqrt{v_t^2 t^2 + y_0^2} \right) \right) \quad (4.21)$$

As in the non-relativistic case the transmitted signal is a complex sinusoidal waveform $s(t) = \exp(j2\pi f_0 t)$.

$$g(t) = \frac{c_0}{\sqrt{c_0^2 - v_t^2}} \left| 1 - \frac{v_t}{c_0} \frac{v_t t}{\sqrt{v_t^2 t^2 + y_0^2}} \right| \cdot \exp\left(\underbrace{ j2\pi f_0 \frac{c_0}{\sqrt{c_0^2 - v_t^2}} \left(t - c_0^{-1}\sqrt{v_t^2 t^2 + y_0^2} \right) }_{\varphi(t)} \right)$$

The instantaneous frequency $f(t)$ of the received signal is calculated using the derivative of its phase $\varphi(t)$ with respect to time.

$$f(t) = \frac{1}{2\pi} \frac{d\varphi(t)}{dt} = f_0 \frac{c_0}{\sqrt{c_0^2 - v_t^2}} \left(1 - \frac{v_t}{c_0} \frac{v_t t}{\sqrt{v_t^2 t^2 + y_0^2}} \right) \quad (4.22)$$

From two limiting cases frequency scaling factors according to eq. 3.17 are obtained.

$$t \to \infty \Rightarrow f(\infty) = f_0 \sqrt{\frac{1 - v_t/c_0}{1 + v_t/c_0}} \qquad t \to -\infty \Rightarrow f(-\infty) = f_0 \sqrt{\frac{1 + v_t/c_0}{1 - v_t/c_0}}$$

Fig. 14 shows some examples of normalised instantaneous frequencies as functions of time. The parameters $y_0 = 100$ km and $c_0 = 300.000$ km/s were used for the numerical calculations.

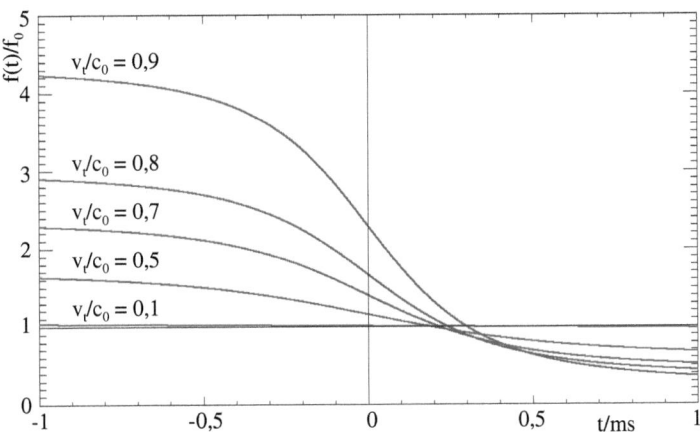

Fig. 14: Instantaneous frequencies of received signals while the transmitter passes a receiver located at the point of origin, relativistic case

In contrast to fig. 13 the curves do not intersect in one point. This indicates that a transverse Doppler Effect occurs in the relativistic case [4]. If the velocity v_t of the transmitter is close to zero, the intersection with the line $f(t) = f_0$ occurs at $t_0 = y_0/2c_0$. At first glance, the deviation from the non-relativistic point of intersection $t_0 = y_0/c_0$ is surprising. However the deviation is easily explained using the symmetry of the relativistic Doppler Effect with respect to motion of the transmitter or the receiver.

In the limiting cases $v_t, v_r \to 0$ t_0 can not tend to y_0/c_0 (non-relativistic case for $v_r = 0$) <u>and</u> to 0 (non-relativistic case for $v_t = 0$) simultaneously.

Replacing the expression $v_t t$ for the x-coordinate of the transmitter by the spatial variable x, the instantaneous frequency is obtained as a function of the spatial position of the transmitter.

$$f(x) = f_0 \frac{c_0}{\sqrt{c_0^2 - v_t^2}} \left(1 - \frac{v_t}{c_0} \frac{x}{\sqrt{x^2 + y_0^2}}\right) \tag{4.23}$$

The expression $x/\sqrt{x^2 + y_0^2}$ is equivalent to the cosine of the angle θ between the line of sight from the receiver to the transmitter and the motion vector of the transmitter according to fig. 10. If $x/\sqrt{x^2 + y_0^2}$ is replaced by $\cos(\theta)$, a modified version of eq. 4.23 is obtained. This version was already mentioned by Einstein in his famous paper "Zur Elektrodynamik bewegter Körper" [2].

In the special case, illustrated in fig. 10, eq. 4.19 simplifies to:

$$t = \frac{c_0}{\sqrt{c_0^2 - v_t^2}} \left(\tau + c_0^{-1}\sqrt{v_t^2 \tau^2 + y_0^2}\right).$$

If the transmission time τ in eq. 4.19 is replaced by x/v_t, another expression for the instantaneous frequency $f(x)$ as a function of the spatial position of the transmitter is obtained from eq. 4.22.

$$f(x) = f_0 \frac{c_0}{\sqrt{c_0^2 - v_t^2}} \left(1 - \frac{v_t}{c_0} \frac{v_t \frac{c_0}{\sqrt{c_0^2 - v_t^2}}\left(\frac{x}{v_t} + \frac{1}{c_0}\sqrt{v_t^2 \frac{x^2}{v_t^2} + y_0^2}\right)}{\sqrt{\left(v_t^2 \left(\frac{c_0}{\sqrt{c_0^2 - v_t^2}}\left(\frac{x}{v_t} + \frac{1}{c_0}\sqrt{v_t^2 \frac{x^2}{v_t^2} + y_0^2}\right)\right)^2 + y_0^2\right)}}\right)$$

Before this equation can be compared with eq. 4.23, some simplifications are required. The details of this calculation can be found in appendix A.1.

$$f(x) = f_0 \frac{\sqrt{1 - v_t^2/c_0^2}}{1 + \frac{v_t}{c_0} \frac{x}{\sqrt{x^2 + y_0^2}}} \tag{4.24}$$

The equation which is obtained if $x/\sqrt{x^2+y_0^2}$ is replaced by $\cos(\theta)$ is mentioned e. g. in [7] and in many other references. Eq. 4.23 is applicable if the coordinate system of the receiver is used. Eq. 4.24 is applicable if the coordinate system of the transmitter is used.

Fig. 15 shows some examples of normalised instantaneous frequencies as functions of the spatial position of the transmitter using $y_0 = 100$ km and $c_0 = 300.000$ km/s.

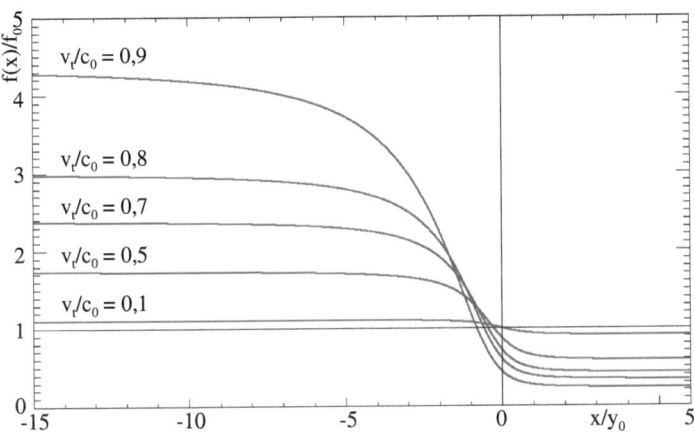

Fig. 15: Instantaneous frequencies of received signals while the transmitter passes a receiver located at the point of origin, relativistic case

4.2.2 Stationary Transmitter

Now the Doppler Effect is investigated for a stationary transmitter and a receiver in uniform motion. Eq. 4.3 is solved for τ. The velocity v_t of the transmitter is set to zero. The abbreviation $\vec{r}_0 = \vec{r}_{r0} - \vec{r}_{t0}$ reduces the length of the equations.

$$(\vec{v}_r t_r + \vec{r}_0) \cdot (\vec{v}_r t_r + \vec{r}_0) = c_0^2 (t_r - \tau)^2 \Rightarrow \tau^2 - 2t_r \tau + \frac{\left(c_0^2 - \vec{v}_r \cdot \vec{v}_r\right) t_r^2 - 2\vec{r}_0 \cdot \vec{v}_r t_r - \vec{r}_0 \cdot \vec{r}_0}{c_0^2} = 0$$

$$\tau = t_r - c_0^{-1} \sqrt{(\vec{v}_r t_r + \vec{r}_0) \cdot (\vec{v}_r t_r + \vec{r}_0)} \qquad (4.25)$$

The Lorentz transform according to eq. 3.15 is applied to the transmission time τ.

$$\tau'(t_r) = \frac{\tau(t_r)}{\sqrt{1 - \vec{v}_r \cdot \vec{v}_r / c_0^2}} = \frac{c_0 t_r - \sqrt{(\vec{v}_r t_r + \vec{r}_0) \cdot (\vec{v}_r t_r + \vec{r}_0)}}{\sqrt{c_0^2 - \vec{v}_r \cdot \vec{v}_r}} \qquad (4.26)$$

The calculation of the time variant impulse response h(t,τ), according to eq. 2.2, is omitted. Using the following expressions in eq. 4.6a simplifies the subsequent calculations.

$$\tau'(t) = \frac{c_0 t - \sqrt{(\vec{v}_r t + \vec{r}_0) \cdot (\vec{v}_r t + \vec{r}_0)}}{\sqrt{c_0^2 - \vec{v}_r \cdot \vec{v}_r}} \Rightarrow \frac{d\tau'(t)}{dt} = \frac{1}{\sqrt{c_0^2 - \vec{v}_r \cdot \vec{v}_r}} \left(c_0 - \frac{(\vec{v}_r t + \vec{r}_0) \cdot \vec{v}_r}{\sqrt{(\vec{v}_r t + \vec{r}_0) \cdot (\vec{v}_r t + \vec{r}_0)}} \right)$$

The received signal g(t) is obtained as:

$$g(t) = \frac{1}{\sqrt{c_0^2 - \vec{v}_r \cdot \vec{v}_r}} \left| c_0 - \frac{(\vec{v}_r t + \vec{r}_0) \cdot \vec{v}_r}{\sqrt{(\vec{v}_r t + \vec{r}_0) \cdot (\vec{v}_r t + \vec{r}_0)}} \right| \cdot s \left(\frac{c_0 t - \sqrt{(\vec{v}_r t + \vec{r}_0) \cdot (\vec{v}_r t + \vec{r}_0)}}{\sqrt{c_0^2 - \vec{v}_r \cdot \vec{v}_r}} \right) \quad (4.27)$$

Without loss of generality the same simplifications as in the non-relativistic case are used. The receiver is moving on a line parallel to the x-axis of the coordinate system, i. e. $\vec{r}_0 = (0, y_0, 0)$ resp. $\vec{v}_r = (v_r, 0, 0)$. Such eq. 4.6 can be replaced by a much simpler equation. Fig. 12 illustrates this configuration.

$$g(t) = \frac{c_0}{\sqrt{c_0^2 - v_r^2}} \left| 1 - \frac{v_r}{c_0} \frac{v_r t}{\sqrt{v_r^2 t^2 + y_0^2}} \right| \cdot s \left(\frac{c_0}{\sqrt{c_0^2 - v_r^2}} \left(t - c_0^{-1} \sqrt{v_r^2 t^2 + y_0^2} \right) \right) \quad (4.28)$$

This equation is equivalent to eq. 4.21. The symmetry of the relativistic Doppler Effect with respect to motion of the transmitter or the receiver confirms the results of these calculations. Such a calculation of the instantaneous frequency for a moving receiver is not needed.

4.2.3 Motion of Transmitter and Receiver

If the transmitter and the receiver are in motion, the Doppler Effect is a function of the relativistic subtraction of velocities of the transmitter and the receiver [4, 6]. The relativistic subtraction of velocities can be written by using vectors.

$$\vec{v}_d = \frac{\vec{v}_t \cdot \vec{v}_r / \vec{v}_r \cdot \vec{v}_r - 1}{1 - \vec{v}_t \cdot \vec{v}_r / c_0^2} \vec{v}_r + \left(\vec{v}_t - \frac{\vec{v}_t \cdot \vec{v}_r}{\vec{v}_r \cdot \vec{v}_r} \vec{v}_r \right) \frac{\sqrt{1 - \vec{v}_r \cdot \vec{v}_r / c_0^2}}{1 - \vec{v}_t \cdot \vec{v}_r / c_0^2} \quad (4.29)$$

The detailed derivation and the discussion of eq. 4.29 can be found in appendix A.2.

The received signal is calculated using eq. 4.20c. The motion vector $\vec{v}_t$ is replaced by the relativistic subtraction of velocities $\vec{v}_d$ according to eq. 4.29.

$$g(t) = \left| \frac{c_0}{\sqrt{c_0^2 - \vec{v}_d \bullet \vec{v}_d}} + \frac{t - \frac{\sqrt{c_0^2 - \vec{v}_d \bullet \vec{v}_d} c_0 t + \vec{r}_0 \bullet \vec{v}_d}{c_0^2 - \vec{v}_d \bullet \vec{v}_d}}{\sqrt{\left(\frac{\sqrt{c_0^2 - \vec{v}_d \bullet \vec{v}_d} c_0 t + \vec{r}_0 \bullet \vec{v}_d}{c_0^2 - \vec{v}_d \bullet \vec{v}_d}\right)^2 - \frac{\left(c_0^2 - \vec{v}_d \bullet \vec{v}_d\right)t^2 - \vec{r}_0 \bullet \vec{r}_0}{c_0^2 - \vec{v}_d \bullet \vec{v}_d}}} \frac{c_0}{\sqrt{c_0^2 - \vec{v}_d \bullet \vec{v}_d}} \right|$$

$$\cdot s\left(\frac{\sqrt{c_0^2 - \vec{v}_d \bullet \vec{v}_d} c_0 t + \vec{r}_0 \bullet \vec{v}_d}{c_0^2 - \vec{v}_d \bullet \vec{v}_d} - \sqrt{\left(\frac{\sqrt{c_0^2 - \vec{v}_d \bullet \vec{v}_d} c_0 t + \vec{r}_0 \bullet \vec{v}_d}{c_0^2 - \vec{v}_d \bullet \vec{v}_d}\right)^2 - \frac{\left(c_0^2 - \vec{v}_d \bullet \vec{v}_d\right)t^2 - \vec{r}_0 \bullet \vec{r}_0}{c_0^2 - \vec{v}_d \bullet \vec{v}_d}} \right)$$

(4.30)

If eq. 4.29 was used in eq. 4.30 an extremely bulky and complicated equation would be obtained which would not provide any further insight.

5 Monostatic Radar

Without loss of generality the locations of the transmitter and the receiver are assumed to be identical. Throughout this chapter the term radar is used for all types of transmitter-receiver configurations including acoustic systems like sonar or optical systems like lidar.

Two different configurations are investigated and compared.

1. Stationary radar and uniform motion of reflector
2. Stationary reflector and uniform motion of radar

The term reflector is used for an isotropic point scatterer.

5.1 Stationary Radar and uniform Motion of Reflector

First a stationary radar located at the point of origin is considered. The reflector moves with constant velocity on a line parallel to the x-axis of the coordinate system, i. e. $\vec{r}_{s0} = (0, y_0, 0)$, $\vec{v} = (v, 0, 0)$. Without loss of generality all z-components are set to zero. Fig. 16 illustrates this configuration.

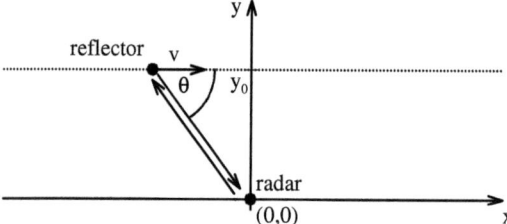

Fig. 16: Reflector passing a radar with constant velocity

5.1.1 Non-relativistic Doppler Effect

The locations of the radar resp. the reflector are given by eq. 5.1.

$$x = 0, \ y = 0 \text{ resp. } x = vt, \ y = y_0 \tag{5.1}$$

At time τ the radar emits a spherical wave. Eq. 5.2 characterises the evolution of the sphere as a function of time.

$$x^2 + y^2 = c_0^2 (t - \tau)^2 \tag{5.2}$$

The squared radius of the sphere at time t is on the right-hand side of the equation. The radius is proportional to the velocity of propagation c_0 of the signal.

Fig. 17 illustrates the situation.

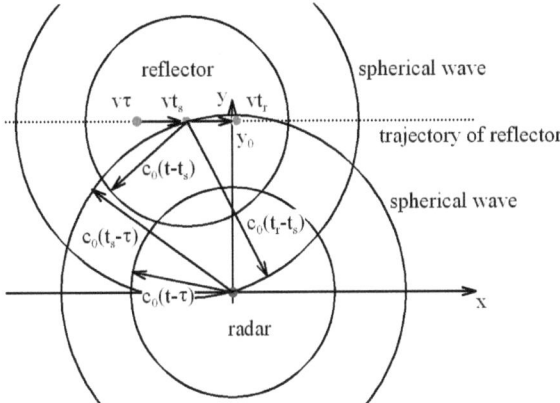

Fig. 17: Propagation of spherical waves from the radar to the reflector and back to the radar

At time t_s the spherical wave arrives at the reflector. This time is calculated using the trajectory of the reflector according to eq. 5.1 in eq. 5.2.

$$v^2 t_s^2 + y_0^2 = c_0^2 (t_s - \tau)^2 \tag{5.3}$$

The scattering time t_s is calculated by solving the following quadratic equation.

$$v^2 t_s^2 + y_0^2 = c_0^2 \tau^2 - 2 c_0^2 t_s \tau + c_0^2 t_s^2 \Rightarrow t_s^2 - 2 \frac{c_0^2 \tau}{c_0^2 - v^2} t_s + \frac{c_0^2 \tau^2 - y_0^2}{c_0^2 - v^2} = 0$$

$$t_s = \frac{c_0^2 \tau}{c_0^2 - v^2} + \sqrt{\left(\frac{c_0^2 \tau}{c_0^2 - v^2}\right)^2 - \frac{c_0^2 \tau^2 - y_0^2}{c_0^2 - v^2}} = \frac{c_0^2 \tau + \sqrt{c_0^4 \tau^2 - (c_0^2 - v^2)(c_0^2 \tau^2 - y_0^2)}}{c_0^2 - v^2}$$

$$t_s = \frac{c_0^2}{c_0^2 - v^2}\left(\tau + c_0^{-1}\sqrt{v^2 \tau^2 + \left(1 - \frac{v^2}{c_0^2}\right) y_0^2}\right) \tag{5.4}$$

The plus sign in front of the square root is selected because of the causality condition $t_s \geq \tau$.

At time t_s scattering of the transmitted spherical wave generates another spherical wave. Eq. 5.5 characterises the evolution of the new sphere as a function of time.

$$(x - v t_s)^2 + (y - y_0)^2 = c_0^2 (t - t_s)^2 \tag{5.5}$$

At time t_r the scattered spherical wave arrives at the radar. This time can be calculated using the location of the radar according to eq. 5.1 in eq. 5.5.

$$v^2 t_s^2 + y_0^2 = c_0^2 (t_r - t_s)^2 \tag{5.6}$$

Now eq. 5.6 is also solved for the scattering time t_s.

$$v^2 t_s^2 + y_0^2 = c_0^2 t_r^2 - 2 c_0^2 t_r t_s + c_0^2 t_s^2 \Rightarrow t_s^2 - 2 \frac{c_0^2 t_r}{c_0^2 - v^2} t_s + \frac{c_0^2 t_r^2 - y_0^2}{c_0^2 - v^2} = 0$$

$$t_s = \frac{c_0^2 t_r}{c_0^2 - v^2} - \sqrt{\left(\frac{c_0^2 t_r}{c_0^2 - v^2}\right)^2 - \frac{c_0^2 t_r^2 - y_0^2}{c_0^2 - v^2}} = \frac{c_0^2 t_r - \sqrt{c_0^2 t_r^2 - (c_0^2 - v^2)(c_0^2 t_r^2 - y_0^2)}}{c_0^2 - v^2}$$

The minus sign in front of the square root is selected because of the causality condition $t_s \leq t_r$.

$$t_s = \frac{c_0^2}{c_0^2 - v^2} \left(t_r - c_0^{-1} \sqrt{v^2 t_r^2 + \left(1 - \frac{v^2}{c_0^2}\right) y_0^2} \right) \tag{5.7}$$

t_s from eq. 5.4 is used in eq. 5.7 and thus eliminated.

$$\tau + c_0^{-1} \sqrt{v^2 \tau^2 + \left(1 - \frac{v^2}{c_0^2}\right) y_0^2} = t_r - c_0^{-1} \sqrt{v^2 t_r^2 + \left(1 - \frac{v^2}{c_0^2}\right) y_0^2} \tag{5.8}$$

Solving eq. 5.8 for t_r enables the calculation of the time variant impulse response $h(t,\tau)$ according to eq. 2.2.

$$t_r = \frac{c_0^2}{c_0^2 - v^2} \left(\tau + \sqrt{\frac{v^2}{c_0^2} \tau^2 + \frac{y_0^2}{c_0^2} \frac{c_0^2 - v^2}{c_0^2}} \right) + \sqrt{\frac{c_0^2 v^2}{(c_0^2 - v^2)^2} \left(\tau + \sqrt{\frac{v^2}{c_0^2} \tau^2 + \frac{y_0^2}{c_0^2} \frac{c_0^2 - v^2}{c_0^2}} \right)^2 + \frac{y_0^2}{c_0^2}} \tag{5.9}$$

The details of this calculation can be found in appendix A.3. The time variant impulse response $h(t,\tau)$, according to eq. 2.2, is obtained as:

$$h(t,\tau) = \delta(t - t_r(\tau)) = \delta \left(t - \frac{c_0^2}{c_0^2 - v^2} \left(\tau + \sqrt{\frac{v^2}{c_0^2} \tau^2 + \frac{y_0^2}{c_0^2} \frac{c_0^2 - v^2}{c_0^2}} \right) \right.$$
$$\left. - \sqrt{\frac{c_0^2 v^2}{(c_0^2 - v^2)^2} \left(\tau + \sqrt{\frac{v^2}{c_0^2} \tau^2 + \frac{y_0^2}{c_0^2} \frac{c_0^2 - v^2}{c_0^2}} \right)^2 + \frac{y_0^2}{c_0^2}} \right) \tag{5.10}$$

Solving eq. 5.8 for τ enables the direct calculation of the received signal g(t).

$$v^2\tau^2 + \left(1-\frac{v^2}{c_0^2}\right)y_0^2 = \left(\left(c_0 t_r - \sqrt{v^2 t_r^2 + \left(1-\frac{v^2}{c_0^2}\right)y_0^2}\right) - c_0\tau\right)^2$$

$$\tau^2 - 2\frac{c_0 t_r - \sqrt{v^2 t_r^2 + \left(1-\frac{v^2}{c_0^2}\right)y_0^2}}{c_0^2 - v^2}c_0\tau + \frac{\left(c_0 t_r - \sqrt{v^2 t_r^2 + \left(1-\frac{v^2}{c_0^2}\right)y_0^2}\right)^2 - \left(1-\frac{v^2}{c_0^2}\right)y_0^2}{c_0^2 - v^2} = 0 \quad (5.11)$$

After replacing the receiving time t_r by t the time of emission $\tau(t)$ and its derivative $d\tau(t)/dt$ are determined by solving the following quadratic equation.

$$\tau = \frac{c_0 t_r - \sqrt{v^2 t_r^2 + \left(1-\frac{v^2}{c_0^2}\right)y_0^2}}{c_0^2 - v^2}c_0$$

$$- \sqrt{\left(\frac{c_0 t_r - \sqrt{v^2 t_r^2 + \left(1-\frac{v^2}{c_0^2}\right)y_0^2}}{c_0^2 - v^2}c_0\right)^2 - \frac{\left(c_0 t_r - \sqrt{v^2 t_r^2 + \left(1-\frac{v^2}{c_0^2}\right)y_0^2}\right)^2 - \left(1-\frac{v^2}{c_0^2}\right)y_0^2}{c_0^2 - v^2}}$$

$$\tau = \frac{c_0^2}{c_0^2 - v^2}\left(t - \sqrt{\frac{v^2}{c_0^2}t^2 + \frac{y_0^2}{c_0^2}\frac{c_0^2 - v^2}{c_0^2}}\right) - \sqrt{\frac{c_0^2 v^2}{(c_0^2 - v^2)^2}\left(t - \sqrt{\frac{v^2}{c_0^2}t^2 + \frac{y_0^2}{c_0^2}\frac{c_0^2 - v^2}{c_0^2}}\right)^2 + \frac{y_0^2}{c_0^2}} \quad (5.12a)$$

The minus sign in front of the square root is selected because of the causality condition $\tau \le t_r$.

$$\frac{d\tau}{dt} = \frac{c_0^2}{c_0^2 - v^2}\left(1 - \frac{\frac{v^2}{c_0^2 - v^2}\left(t - \sqrt{\frac{v^2}{c_0^2}t^2 + \frac{y_0^2}{c_0^2}\frac{c_0^2 - v^2}{c_0^2}}\right)}{\sqrt{\frac{c_0^2 v^2}{(c_0^2 - v^2)^2}\left(t - \sqrt{\frac{v^2}{c_0^2}t^2 + \frac{y_0^2}{c_0^2}\frac{c_0^2 - v^2}{c_0^2}}\right)^2 + \frac{y_0^2}{c_0^2}}}\left(1 - \frac{\frac{v^2}{c_0^2}t}{\sqrt{\frac{v^2}{c_0^2}t^2 + \frac{y_0^2}{c_0^2}\frac{c_0^2 - v^2}{c_0^2}}}\right)\right)$$

(5.12b)

Combining eqs. 5.12a and 5.12b the received signal g(t) is obtained.

$$g(t) = \left| \frac{c_0^2}{c_0^2 - v^2} \left(1 - \frac{\frac{v^2}{c_0^2 - v^2}\left(t - \sqrt{\frac{v^2}{c_0^2}t^2 + \frac{y_0^2}{c_0^2}\frac{c_0^2 - v^2}{c_0^2}}\right)}{\sqrt{\frac{c_0^2 v^2}{(c_0^2 - v^2)^2}\left(t - \sqrt{\frac{v^2}{c_0^2}t^2 + \frac{y_0^2}{c_0^2}\frac{c_0^2 - v^2}{c_0^2}}\right)^2 + \frac{y_0^2}{c_0^2}}} \right) \left(1 - \frac{\frac{v^2}{c_0^2}t}{\sqrt{\frac{v^2}{c_0^2}t^2 + \frac{y_0^2}{c_0^2}\frac{c_0^2 - v^2}{c_0^2}}} \right) \right.$$

$$\left. \cdot s\left(\frac{c_0^2}{c_0^2 - v^2}\left(t - \sqrt{\frac{v^2}{c_0^2}t^2 + \frac{y_0^2}{c_0^2}\frac{c_0^2 - v^2}{c_0^2}}\right) - \sqrt{\frac{c_0^2 v^2}{(c_0^2 - v^2)^2}\left(t - \sqrt{\frac{v^2}{c_0^2}t^2 + \frac{y_0^2}{c_0^2}\frac{c_0^2 - v^2}{c_0^2}}\right)^2 + \frac{y_0^2}{c_0^2}} \right) \right|$$

(5.12c)

Now the transmitted signal is a complex sinusoidal waveform $s(t) = \exp(j2\pi f_0 t)$. Eq. 5.12c enables the calculation of the instantaneous frequency of the received signal.

$$g(t) = \left|\frac{d\tau(t)}{dt}\right| \exp\left(j2\pi f_0 \underbrace{\left[\frac{c_0^2}{c_0^2 - v^2}\left(t - \sqrt{\frac{v^2}{c_0^2}t^2 + \frac{c_0^2 - v^2}{c_0^2}\frac{y_0^2}{c_0^2}}\right) - \sqrt{\frac{c_0^2 v^2}{(c_0^2 - v^2)^2}\left(t - \sqrt{\frac{v^2}{c_0^2}t^2 + \frac{c_0^2 - v^2}{c_0^2}\frac{y_0^2}{c_0^2}}\right)^2 + \frac{y_0^2}{c_0^2}}\right]}_{\varphi(t)} \right)$$

(5.13a)

Eq. 5.12b is used in eq. 5.13b in order to calculate f(t).

$$f(t) = \frac{1}{2\pi}\frac{d\varphi(t)}{dt} = \frac{1}{2\pi}2\pi f_0 \frac{d}{dt}\tau(t) \Rightarrow \frac{f(t)}{f_0} = \frac{d}{dt}\tau(t) \qquad (5.13b)$$

For the special case $t \to \infty$ a constant instantaneous frequency $f(\infty) = f_0(1-v/c_0)/(1+v/c_0)$ is obtained. The signal amplitude is equal to $|(1-v/c_0)/(1+v/c_0)|$. For the special case $t \to -\infty$ the constant instantaneous frequency is $f(-\infty) = f_0(1+v/c_0)/(1-v/c_0)$ and the signal amplitude is equal to $|f_0(1+v/c_0)/(1-v/c_0)|$.

Fig. 18 shows some examples of normalised instantaneous frequencies as functions of the spatial position of the reflector using $y_0 = 100$ km and $c_0 = 300.000$ km/s. The expression vt for the x-coordinate of the reflector in eq. 5.12b is replaced by the spatial variable x.

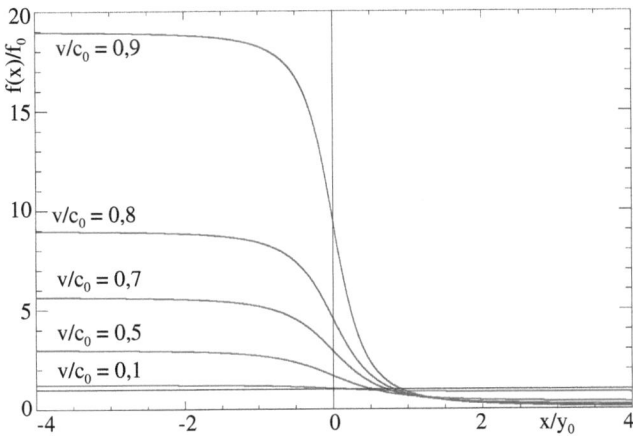

Fig. 18: Instantaneous frequencies of received signals while the reflector passes a radar located at the point of origin, non-relativistic case

5.1.2 Relativistic Doppler Effect

Eq. 5.12a is the starting point for the relativistic calculations. The Lorentz transform is applied to τ and t.

$$\tau = \tau'\sqrt{1-v^2/c_0^2} \quad \text{resp.} \quad t = t'\sqrt{1-v^2/c_0^2} \tag{5.14}$$

τ resp. t in eq. 5.12a are replaced by functions of τ' resp. t' according to eq. 5.14. Then the modified eq. 5.12a is solved for τ'.

$$\tau' = \frac{c_0^2}{c_0^2 - v^2}\left(\left(1+\frac{v^2}{c_0^2}\right)t' - 2\sqrt{\frac{v^2}{c_0^2}t'^2 + \frac{y_0^2}{c_0^2}}\right) \tag{5.15a}$$

The details of this calculation can be found in appendix A.4. After replacing t' by t in eq. 5.15a the signal amplitude is calculated using:

$$\frac{d\tau'}{dt} = \frac{c_0^2}{c_0^2-v^2}\left(\frac{d}{dt}\left(1+\frac{v^2}{c_0^2}\right)t - 2\frac{d}{dt}\sqrt{\frac{v^2}{c_0^2}t^2 + \frac{y_0^2}{c_0^2}}\right) = \frac{c_0^2}{c_0^2-v^2}\left(1+\frac{v^2}{c_0^2} - 2\frac{v^2}{c_0^2}\frac{t}{\sqrt{\frac{v^2}{c_0^2}t^2 + \frac{y_0^2}{c_0^2}}}\right) \tag{5.15b}$$

Combining eqs. 5.15a and 5.15b the received signal g(t) is obtained.

$$g(t) = \left| \frac{1}{c_0^2 - v^2} \left(c_0^2 - 2v^2 \frac{c_0 t}{\sqrt{v^2 t^2 + y_0^2}} + v^2 \right) \right| \cdot s\left(\frac{c_0^2}{c_0^2 - v^2} \left(\left(1 + \frac{v^2}{c_0^2}\right) t - 2\sqrt{\frac{v^2}{c_0^2} t^2 + \frac{y_0^2}{c_0^2}} \right) \right) \quad (5.15c)$$

The instantaneous frequency f(t) of the received signal g(t) is calculated using the complex sinusoidal waveform s(t) = exp(j2πf₀t) in eq. 5.15c.

$$g(t) = \left| \frac{d\tau(t)}{dt} \right| \cdot \exp\left(j2\pi f_0 \underbrace{\frac{c_0^2}{c_0^2 - v^2} \left(\left(1 + \frac{v^2}{c_0^2}\right) t - 2\sqrt{\frac{v^2}{c_0^2} t^2 + \frac{y_0^2}{c_0^2}} \right)}_{\varphi(t)} \right) \quad (5.16a)$$

Eq. 5.15b is used in eq. 5.16b in order to calculate f(t).

$$\frac{f(t)}{f_0} = \frac{d}{dt}\tau'(t) = \frac{1}{1 - v^2/c_0^2}\left(1 - 2\frac{v}{c_0}\frac{vt}{\sqrt{v^2 t^2 + y_0^2}} + \frac{v^2}{c_0^2} \right) \quad (5.16b)$$

For the special case $t \to \infty$ a constant instantaneous frequency $f(\infty) = f_0(1-v/c_0)/(1+v/c_0)$ is obtained. The signal amplitude is equal to $|(1-v/c_0)/(1+v/c_0)|$. For the special case $t \to -\infty$ the constant instantaneous frequency is $f(-\infty) = f_0(1+v/c_0)/(1-v/c_0)$ and the signal amplitude is equal to $|f_0(1+v/c_0)/(1-v/c_0)|$. The special cases are identical to the special cases of the non-relativistic equation 5.13a.

If the expression vt for the x-coordinate of the reflector is replaced by the spatial variable x eq. 5.16b is replaced by:

$$\frac{f(x)}{f_0} = \frac{1}{1 - v^2/c_0^2}\left(1 - 2\frac{v}{c_0}\frac{x}{\sqrt{x^2 + y_0^2}} + \frac{v^2}{c_0^2} \right) \quad (5.17)$$

The expression $x/\sqrt{x^2 + y_0^2}$ is equivalent to the cosine of the angle θ between the line of sight from the radar to the reflector and the motion vector of the reflector according to fig. 16. If $x/\sqrt{x^2 + y_0^2}$ is replaced by cos(θ) a modified version of eq. 5.17 is obtained. This version was already mentioned by Einstein in his famous paper "Zur Elektrodynamik bewegter Körper" [2]. Eq. 5.15c is also confirmed by the amplitude scaling factor mentioned by Einstein.

Fig. 19 shows some examples of normalised instantaneous frequencies as functions of the spatial position of the reflector according to eq. 5.17 using $y_0 = 100$ km and $c_0 = 300.000$ km/s.

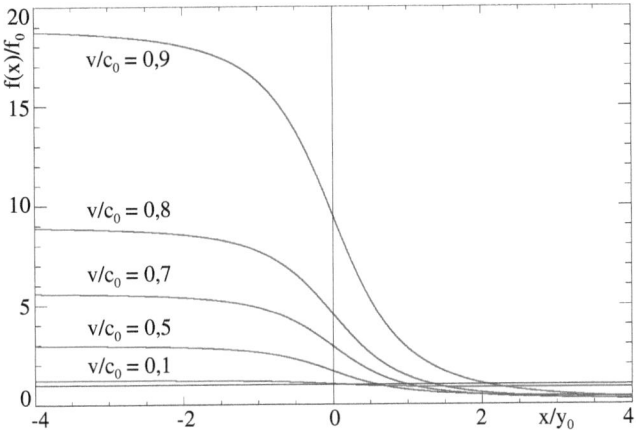

Fig. 19: Instantaneous frequencies of received signals while the reflector passes a radar located at the point of origin, relativistic case

5.2 Stationary Reflector and uniform Motion of Radar

Now a stationary reflector located at the point of origin is considered. The radar moves with constant velocity on a line parallel to the x-axis of the coordinate system, i. e. $\vec{r}_{r0} = (0, y_0, 0)$, $\vec{v} = (v, 0, 0)$. Without loss of generality all z-components are set to zero. Fig. 20 illustrates this configuration.

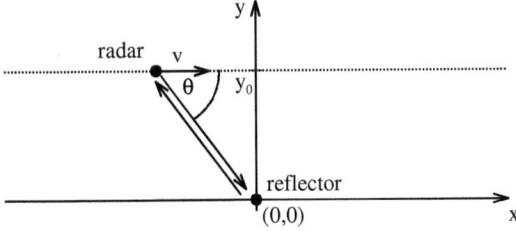

Fig. 20: Radar passing a reflector with constant velocity

The locations of the radar resp. the reflector are given by eq. 5.18.

$$x = vt, \; y = y_0 \; \text{resp.} \; x = 0, \; y = 0 \qquad (5.18)$$

At time τ the radar emits a spherical wave. Eq. 5.19 characterises the evolution of the sphere as a function of time.

$$(x-v\tau)^2 + (y-y_0)^2 = c_0^2(t-\tau)^2 \tag{5.19}$$

The squared radius of the sphere at time t is on the right-hand side of the equation. The radius is proportional to the velocity of propagation c_0 of the signal.

At time t_s the spherical wave arrives at the reflector. This time is calculated using the location of the reflector according to eq. 5.18 in eq. 5.19.

$$v^2\tau^2 + y_0^2 = c_0^2(t_s - \tau)^2 \tag{5.20}$$

The scattering time t_s is calculated by solving this quadratic equation.

$$t_s = \tau + c_0^{-1}\sqrt{v^2\tau^2 + y_0^2} \tag{5.21}$$

The plus sign in front of the square root is selected because of the causality condition $t_s \geq \tau$. At time t_s scattering of the transmitted spherical wave generates another spherical wave. Eq. 5.22 characterises the evolution of the new sphere as a function of time.

$$x^2 + y^2 = c_0^2(t-t_s)^2 \tag{5.22}$$

At time t_r the scattered spherical wave arrives at the radar. This time can be calculated using the trajectory of the radar according to eq. 5.18 in eq. 5.22.

$$v^2 t_r^2 + y_0^2 = c_0^2(t_r - t_s)^2 \tag{5.23}$$

Now eq. 5.23 is also solved for the scattering time t_s.

$$t_s = t_r - c_0^{-1}\sqrt{v^2 t_r^2 + y_0^2} \tag{5.24}$$

The minus sign in front of the square root is selected because of the causality condition $t_s \leq t_r$. t_s from eq. 5.21 is used in eq. 5.24 and thus eliminated.

$$\tau + c_0^{-1}\sqrt{v^2\tau^2 + y_0^2} = t_r - c_0^{-1}\sqrt{v^2 t_r^2 + y_0^2}$$

Solving this equation for τ enables the direct calculation of the received signal g(t) according to eq. 4.6a.

$$\tau = \frac{c_0^2}{c_0^2 - v^2}\left(\left(1 + \frac{v^2}{c_0^2}\right)t_r - 2\sqrt{\frac{v^2}{c_0^2}t_r^2 + \frac{y_0^2}{c_0^2}}\right) \tag{5.25}$$

This equation is identical to eq. 5.15a. The details of this calculation can be found in appendix A.5.

If the radar is in uniform motion and the reflector is stationary the results of the non-relativistic and the relativistic calculations are identical. At first glance this identity seems to contradict intuition. However there is a simple explanation of this result. If the reflector is stationary the Lorentz transform must be applied to the scattering time t_s according to eq. 5.21 and to eq. 5.24. If eq. 5.21 is used in eq. 5.24 the Lorentz transforms cancel. The symmetry of the relativistic Doppler Effect with respect to motion of the transmitter or the receiver confirms the identity of eqs. 5.15a and 5.25.

5.3 Airborne Radar over Ground

Fig. 21 shows a radar-reflector constellation which occurs very often in Remote Sensing. An airborne or satellite based radar moves with constant velocity parallel to the ground. For simplicity the earth's surface is replaced by a plane surface. On a local scale the accuracy of this approximation is sufficient for practical applications. h denotes the flight altitude of the radar. The radar target is a single point reflector on the surface. A more realistic received signal can be obtained by integration over the surface using spatially varying reflection coefficients $\rho(x_s,y_s)$.

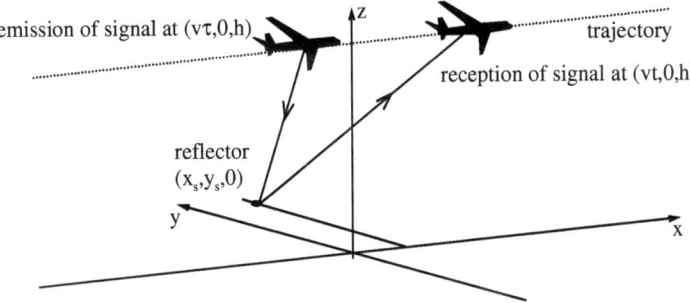

Fig. 21: Airborne radar over the earth's surface

The locations of the radar resp. the reflector are given by eq. 5.26.

$$x = v t, \; y = 0, z = h \; \text{resp.} \; x = x_s, \; y = y_s, z = 0 \tag{5.26}$$

At time τ the radar emits a spherical wave. Eq. 5.27 characterises the evolution of the sphere as a function of time.

$$(x - v\tau)^2 + y^2 + (z - h)^2 = c_0^2 (t - \tau)^2 \tag{5.27}$$

The squared radius of the sphere at time t is on the right-hand side of the equation. The radius is proportional to the velocity of propagation c_0 of the signal. At time t_s the spherical wave arrives at the reflector. This time is calculated using the location of the reflector according to eq. 5.26 in eq. 5.27.

$$(x_s - v\tau)^2 + y_s^2 + h^2 = c_0^2 (t_s - \tau)^2 \tag{5.28}$$

The scattering time t_s is calculated by solving this quadratic equation.

$$t_s = \tau + c_0^{-1}\sqrt{(x_s - v\tau)^2 + y_s^2 + h^2} \qquad (5.29)$$

The plus sign in front of the square root is selected because of the causality condition $t_s \geq \tau$.

At time t_s scattering of the transmitted spherical wave generates another spherical wave. Eq. 5.30 characterises the evolution of the new sphere as a function of time.

$$(x - x_s)^2 + (y - y_s)^2 + z^2 = c_0^2(t - t_s)^2 \qquad (5.30)$$

At time t_r the scattered spherical wave arrives at the radar. This time can be calculated using the trajectory of the radar according to eq. 5.26 in eq. 5.30.

$$(x_s - vt_r)^2 + y_s^2 + h^2 = c_0^2(t_r - t_s)^2 \qquad (5.31)$$

Now eq. 5.31 is also solved for the scattering time t_s.

$$t_s = t_r - c_0^{-1}\sqrt{(x_s - vt_r)^2 + y_s^2 + h^2} \qquad (5.32)$$

The minus sign in front of the square root is selected because of the causality condition $t_s \leq t_r$.

5.3.1 Calculation of the received Signal in General Form

t_s from eq. 5.29 is used in eq. 5.32 and thus eliminated. Solving this equation for τ enables the direct calculation of the received signal g(t) according to eq. 4.6a. The receiving time t_r is replaced by t.

$$\tau = \frac{c_0^2 t - c_0\sqrt{(x_s - vt)^2 + y_s^2 + h^2} - x_s v}{c_0^2 - v^2} - \sqrt{\left(\frac{c_0^2 t - c_0\sqrt{(x_s - vt)^2 + y_s^2 + h^2} - x_s v}{c_0^2 - v^2}\right)^2 - \frac{\left(c_0 t - \sqrt{(x_s - vt)^2 + y_s^2 + h^2}\right)^2 - x_s^2 - y_s^2 - h^2}{c_0^2 - v^2}}$$

(5.33a)

The details of this calculation can be found in appendix A.6.

The signal amplitude is calculated using:

$$\frac{d\tau}{dt} = \frac{c_0^2}{c_0^2 - v^2}\left(1 + \frac{v}{c_0}\frac{x_s - vt}{\sqrt{(x_s - vt)^2 + y_s^2 + h^2}}\right) +$$

$$\frac{\frac{c_0^2}{c_0^2 - v^2}\left(1 + \frac{v}{c_0}\frac{x_s - vt}{\sqrt{(x_s - vt)^2 + y_s^2 + h^2}}\right)\frac{v}{c_0}\left(c_0 x_s - v\left(c_0 t - \sqrt{(x_s - vt)^2 + y_s^2 + h^2}\right)\right)}{\sqrt{\left(c_0^2 t - c_0\sqrt{(x_s - vt)^2 + y_s^2 + h^2} - x_s v\right)^2 - (c_0^2 - v^2)\left(\left(c_0 t - \sqrt{(x_s - vt)^2 + y_s^2 + h^2}\right)^2 - x_s^2 - y_s^2 - h^2\right)}}$$

(5.33b)

The details of this calculation can also be found in appendix A.6. Combing eqs. 5.33a and 5.33b according to eq. 5.33c resp. eq. 4.6a the received signal g(t) could be calculated in principle.

$$g(t) = \left|\frac{d\tau(t)}{dt}\right| \cdot s(\tau(t))$$

(5.33c)

However this solution would be very large and not very clear. Since no substantial simplifications are possible the detailed notation of this bulky equation is omitted.

Fig. 22 shows some examples of normalised instantaneous frequencies as functions of time using $f_0 = 1$ GHz, $h = 5$ km, $x_s = 5$ km, $y_s = 10$ km, $v = 1000$ km/h, 2000 km/h resp. 3000 km/h. The instantaneous frequencies are calculated according to eqs. 5.13b and 5.33b.

$$f(t) = f_0 \cdot \frac{d\tau(t)}{dt}$$

(5.34)

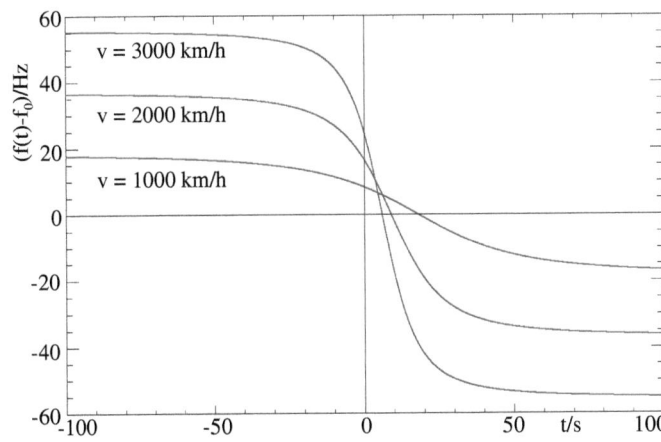

Fig. 22: Instantaneous frequencies of received signals as functions of time

Fig. 23 shows the normalised instantaneous frequencies as functions of the spatial position x = vt of the radar using the same parameters as in fig. 22.

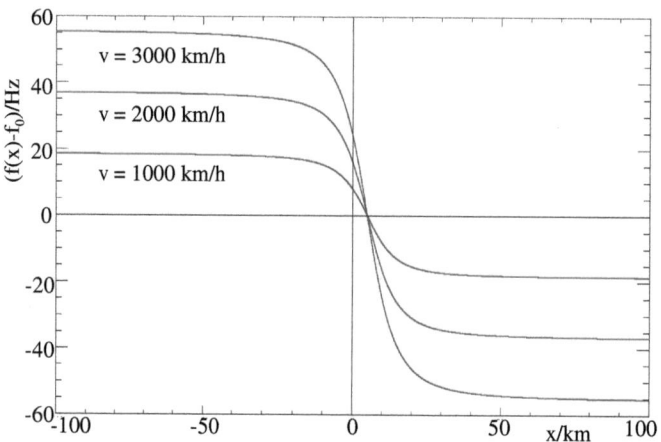

Fig. 23: Instantaneous frequencies of received signals as functions of spatial position

Since for a stationary reflector the results of the non-relativistic and the relativistic calculations are identical further investigations of the radar-reflector configuration shown in fig. 21 can be omitted.

5.3.2 Application to Synthetic Aperture Radar

The phase difference $\Delta\varphi(t)$ between the received signal g(t) and the transmitted signal s(t) is an important signal parameter e. g. for Synthetic Aperture Radar (SAR) [9]. The transmitted signal is a complex sinusoidal waveform s(t) = exp(j2πf$_0$t) with frequency f$_0$.

$$\Delta\varphi(t) = 2\pi f_0 \tau(t) - 2\pi f_0 t = 2\pi f_0 \left(\tau(t) - t \right)$$

Fig. 24 shows some examples of phase differences using the same parameters as in figs. 22 and 23.

For SAR signal processing the phase differences which are functions of time are approximated by parabolic functions according to eq. 5.35a. The dotted curves in fig. 24 will be derived in the remaining part of chapter 5.

$$\Delta\varphi(t) \approx a + b \cdot (t - t_0) + c \cdot (t - t_0)^2 \tag{5.35a}$$

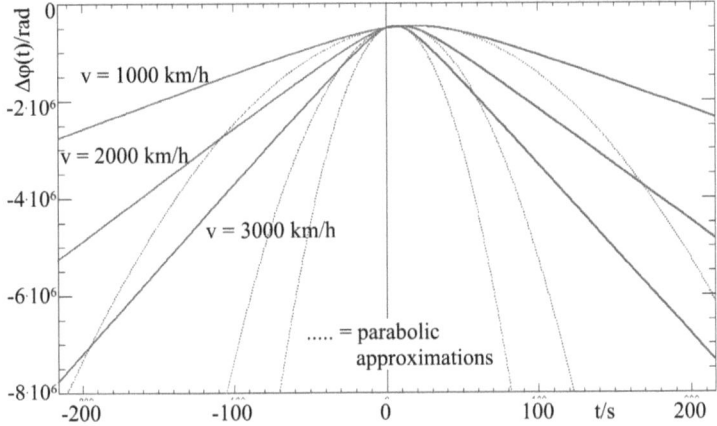

Fig. 24: Phase differences between received signals and transmitted signals

The coefficients a, b and c are derived using a second order Taylor expansion of $\Delta\varphi(t)$ about time t_0.

$$a = \Delta\varphi(t_0) \qquad b = \left.\frac{d\Delta\varphi(t)}{dt}\right|_{t=t_0} \qquad c = \frac{1}{2}\left.\frac{d^2\Delta\varphi(t)}{dt^2}\right|_{t=t_0} \qquad (5.35b)$$

The peak value of $\Delta\varphi(t)$ occurs at time t_0.

$$\left.\frac{d\Delta\varphi(t)}{dt}\right|_{t=t_0} = 2\pi f_0 \left(\left.\frac{d\tau(t)}{dt}\right|_{t=t_0} - 1\right) = 0 \Rightarrow \left.\frac{d\tau(t)}{dt}\right|_{t=t_0} = 1$$

Therefore the coefficient b is equal to zero. t_0 is calculated using eq. 5.33b.

$$1 - \frac{c_0^2}{c_0^2 - v^2}\left(1 + \frac{v}{c_0}\frac{x_s - vt_0}{\sqrt{(x_s - vt_0)^2 + y_s^2 + h^2}}\right) =$$

$$\frac{\frac{c_0^2}{c_0^2 - v^2}\left(1 + \frac{v}{c_0}\frac{x_s - vt_0}{\sqrt{(x_s - vt_0)^2 + y_s^2 + h^2}}\right)\frac{v}{c_0}\left(c_0 x_s - v\left(c_0 t_0 - \sqrt{(x_s - vt_0)^2 + y_s^2 + h^2}\right)\right)}{\sqrt{\left(c_0^2 t_0 - c_0\sqrt{(x_s - vt_0)^2 + y_s^2 + h^2} - x_s v\right)^2 - (c_0^2 - v^2)\left(\left(c_0 t_0 - \sqrt{(x_s - vt_0)^2 + y_s^2 + h^2}\right)^2 - x_s^2 - y_s^2 - h^2\right)}}$$

The result of the lengthy calculation in appendix A.7 is remarkably simple.

$$t_0 = x_s/v \qquad (5.36a)$$

Using eq. 5.36a in eq. 5.33a the coefficient a of the Taylor expansion according to eq. 5.35a is obtained.

$$\tau(t_0) - t_0 = \frac{c_0^2 \frac{x_s}{v} - c_0 \sqrt{\left(x_s - v\frac{x_s}{v}\right)^2 + y_s^2 + h^2} - x_s v}{c_0^2 - v^2} - \frac{x_s}{v}$$

$$\sqrt{\left(\frac{c_0^2 \frac{x_s}{v} - c_0 \sqrt{\left(x_s - v\frac{x_s}{v}\right)^2 + y_s^2 + h^2} - x_s v}{c_0^2 - v^2}\right)^2 - \frac{\left(c_0 \frac{x_s}{v} - \sqrt{\left(x_s - v\frac{x_s}{v}\right)^2 + y_s^2 + h^2}\right)^2 - x_s^2 - y_s^2 - h^2}{c_0^2 - v^2}}$$

The solution of this equation is also remarkably simple. The details of this calculation can be found in appendix A.7.

$$\tau(t_0) - t_0 = -\frac{2c_0}{c_0^2 - v^2} \sqrt{y_s^2 + h^2} \Rightarrow a = -4\pi f_0 \frac{c_0}{c_0^2 - v^2} \sqrt{y_s^2 + h^2} \qquad (5.36b)$$

For the calculation of the coefficient c of the Taylor expansion according to eq. 5.35a the calculation of the second derivative of the phase difference $\Delta\varphi(t)$ resp. the transmission time $\tau(t)$ is required.

$$\frac{d^2\tau}{dt^2} = \frac{d}{dt} \frac{c_0^2}{c_0^2 - v^2} \left(1 + \frac{v}{c_0} \frac{x_s - vt}{\sqrt{(x_s - vt)^2 + y_s^2 + h^2}}\right) +$$

$$\frac{d}{dt} \frac{\frac{c_0^2}{c_0^2 - v^2} \left(1 + \frac{v}{c_0} \frac{x_s - vt}{\sqrt{(x_s - vt)^2 + y_s^2 + h^2}}\right) \frac{v}{c_0} \left(c_0 x_s - v\left(c_0 t - \sqrt{(x_s - vt)^2 + y_s^2 + h^2}\right)\right)}{\sqrt{\left(c_0^2 t - c_0 \sqrt{(x_s - vt)^2 + y_s^2 + h^2} - x_s v\right)^2 - (c_0^2 - v^2)\left(\left(c_0 t - \sqrt{(x_s - vt)^2 + y_s^2 + h^2}\right)^2 - x_s^2 - y_s^2 - h^2\right)}}$$

The solution of this equation is again remarkably simple. The details of this calculation can be found in appendix A.7.

$$\frac{d^2\tau}{dt^2} = -2 \frac{v^2}{\sqrt{y_s^2 + h^2}} \frac{c_0}{c_0^2 - v^2} \Rightarrow c = -\frac{2\pi f_0}{\sqrt{y_s^2 + h^2}} \frac{c_0 v^2}{c_0^2 - v^2} \qquad (5.36c)$$

Combining eqs. 5.36a, 5.36b and 5.36c the phase difference $\Delta\varphi(t)$ is obtained.

$$\Delta\varphi(t) \approx -4\pi f_0 \sqrt{y_s^2 + h^2} \frac{c_0}{c_0^2 - v^2} \left(1 + \frac{1}{2} \frac{v^2}{y_s^2 + h^2} \left(t - \frac{x_s}{v}\right)^2\right)$$

This equation was used for drawing the dotted curves in fig. 24.

6 Non-uniform Motion

Strictly speaking the various types of uniform motion which were considered in the previous chapters are just idealised special cases. Most types of natural motion are non-uniform, e. g. motion of astronomical objects.

6.1 Stationary Receiver and non-uniform Motion of Transmitter

First a stationary receiver located at the point of origin is considered. The transmitter moves along a trajectory $\vec{r}_t(t) = (x_t(t), y_t(t), z_t(t))$. At time τ the transmitter emits a spherical wave. Eq. 6.1 characterises the evolution of the sphere as a function of time.

$$(\vec{r} - \vec{r}_t(\tau)) \cdot (\vec{r} - \vec{r}_t(\tau)) = c_0^2(t-\tau)^2 \tag{6.1}$$

The squared radius of the sphere at time t is on the right-hand side of the equation. The radius is proportional to the velocity of propagation c_0 of the signal.

At time t_r the spherical wave arrives at the receiver. This time is calculated using the location (0, 0, 0) of the receiver in eq. 6.1.

$$\vec{r}_t(\tau) \cdot \vec{r}_t(\tau) = c_0^2(t_r - \tau)^2 \Rightarrow \vec{r}_t(\tau) \cdot \vec{r}_t(\tau) = c_0^2 t_r^2 - 2c_0^2 \tau t_r + c_0^2 \tau^2$$

6.1.1 Non-relativistic Doppler Effect

$$t_r^2 - 2\tau t_r + \frac{c_0^2 \tau^2 - \vec{r}_t(\tau) \cdot \vec{r}_t(\tau)}{c_0^2} = 0 \Rightarrow t_r = \tau + c_0^{-1}\sqrt{\vec{r}_t(\tau) \cdot \vec{r}_t(\tau)} \tag{6.2}$$

The plus sign in front of the square root is selected because of the causality condition $t_r \geq \tau$. The linear superposition according to eq. 2.2 is obtained for arbitrary trajectories of the transmitter. The delta function in the integral provides the time variant impulse response h(t,τ) according to eq. 2.2.

$$g(t) = \int_{-\infty}^{\infty} s(\tau) \delta(t - t_r(\tau)) d\tau = \int_{-\infty}^{\infty} s(\tau) \delta\underbrace{\left(t - \tau - c_0^{-1}\sqrt{\vec{r}_t(\tau) \cdot \vec{r}_t(\tau)}\right)}_{w} d\tau \tag{6.3}$$

A closed form solution for the received signal g(t) is only obtained if the expression for the new variable w can be solved for τ. Otherwise approximations may be found e. g. by linearisation of the argument of the delta function in eq. 6.3 about the time of expansion τ_0.

$$\tau + c_0^{-1}\sqrt{\vec{r}_1(\tau)\cdot\vec{r}_1(\tau)} \approx \tau_0 + c_0^{-1}\sqrt{\vec{r}_1(\tau_0)\cdot\vec{r}_1(\tau_0)} + (\tau-\tau_0)\cdot\frac{d}{d\tau}\left(\tau + c_0^{-1}\sqrt{\vec{r}_1(\tau)\cdot\vec{r}_1(\tau)}\right)\bigg|_{\tau=\tau_0}$$
$$= \tau_0 + c_0^{-1}\sqrt{\vec{r}_1(\tau_0)\cdot\vec{r}_1(\tau_0)} + \left(1 + c_0^{-1}\frac{\vec{r}_1(\tau_0)\cdot\dot{\vec{r}}_1(\tau_0)}{\sqrt{\vec{r}_1(\tau_0)\cdot\vec{r}_1(\tau_0)}}\right)(\tau-\tau_0) \qquad (6.4)$$

Using this expression in eq. 6.3 the expression for w in the argument of the delta function can be solved for τ.

$$t - w \approx \tau_0 + c_0^{-1}\sqrt{\vec{r}_1(\tau_0)\cdot\vec{r}_1(\tau_0)} + \left(1 + c_0^{-1}\frac{\vec{r}_1(\tau_0)\cdot\dot{\vec{r}}_1(\tau_0)}{\sqrt{\vec{r}_1(\tau_0)\cdot\vec{r}_1(\tau_0)}}\right)(\tau-\tau_0) \qquad (6.5)$$

$$\Rightarrow \tau \approx \tau_0 + \frac{c_0(t-w-\tau_0)\sqrt{\vec{r}_1(\tau_0)\cdot\vec{r}_1(\tau_0)} - \vec{r}_1(\tau_0)\cdot\vec{r}_1(\tau_0)}{c_0\sqrt{\vec{r}_1(\tau_0)\cdot\vec{r}_1(\tau_0)} + \vec{r}_1(\tau_0)\cdot\dot{\vec{r}}_1(\tau_0)} \qquad (6.6a)$$

$$\Rightarrow \frac{d\tau}{dw} \approx -\frac{c_0\sqrt{\vec{r}_1(\tau_0)\cdot\vec{r}_1(\tau_0)}}{c_0\sqrt{\vec{r}_1(\tau_0)\cdot\vec{r}_1(\tau_0)} + \vec{r}_1(\tau_0)\cdot\dot{\vec{r}}_1(\tau_0)} \qquad (6.6b)$$

According to eq. 4.6a the received signal g(t) can be expressed as a function of time t and the point of expansion τ_0.

$$g(t) \approx \left|\frac{c_0\sqrt{\vec{r}_1(\tau_0)\cdot\vec{r}_1(\tau_0)}}{c_0\sqrt{\vec{r}_1(\tau_0)\cdot\vec{r}_1(\tau_0)} + \vec{r}_1(\tau_0)\cdot\dot{\vec{r}}_1(\tau_0)}\right|\cdot s\left(\tau_0 + \frac{c_0(t-\tau_0)\sqrt{\vec{r}_1(\tau_0)\cdot\vec{r}_1(\tau_0)} - \vec{r}_1(\tau_0)\cdot\vec{r}_1(\tau_0)}{c_0\sqrt{\vec{r}_1(\tau_0)\cdot\vec{r}_1(\tau_0)} + \vec{r}_1(\tau_0)\cdot\dot{\vec{r}}_1(\tau_0)}\right) \qquad (6.6c)$$

Now the transmitted signal is a complex sinusoidal waveform $s(t) = \exp(j2\pi f_0 t)$. The instantaneous frequency of the received signal is calculated according to eq. 5.13b. After taking the derivative time t no longer occurs and the instantaneous frequency is a function of the point of expansion only.

$$\frac{f(\tau_0)}{f_0} = -\frac{d\tau(w)}{dw} = \frac{c_0\sqrt{\vec{r}_1(\tau_0)\cdot\vec{r}_1(\tau_0)}}{c_0\sqrt{\vec{r}_1(\tau_0)\cdot\vec{r}_1(\tau_0)} + \vec{r}_1(\tau_0)\cdot\dot{\vec{r}}_1(\tau_0)} \qquad (6.7)$$

6.1.2 Relativistic Doppler Effect

In general the derivation of the relativistic Doppler Effect for non-uniform motion requires the application of general relativity [8]. However for moderate accelerations the accuracy of much simpler special relativity [2] is sufficient.

The Lorentz transform is applied to the receiving time t_r of the signal according to eq. 6.2 using the instantaneous velocity $\dot{\vec{r}}_t(\tau)$ of the transmitter at time τ.

$$t'_r = \frac{1}{\sqrt{1-v_t^2(\tau)/c_0^2}}\left(\tau + c_0^{-1}\sqrt{\vec{r}_t(\tau)\cdot\vec{r}_t(\tau)}\right) = \frac{1}{\sqrt{1-\dot{\vec{r}}_t(\tau)\cdot\dot{\vec{r}}_t(\tau)/c_0^2}}\left(\tau + c_0^{-1}\sqrt{\vec{r}_t(\tau)\cdot\vec{r}_t(\tau)}\right) \qquad (6.8)$$

The linear superposition according to eq. 2.2 is obtained for arbitrary trajectories of the transmitter. The delta function in the integral provides the time variant impulse response $h(t,\tau)$ according to eq. 2.2.

$$g(t) = \int_{-\infty}^{\infty} s(\tau)\delta(t-t'_r(\tau))d\tau = \int_{-\infty}^{\infty} s(\tau)\delta\underbrace{\left(t - \frac{c_0\tau + \sqrt{\vec{r}_t(\tau)\cdot\vec{r}_t(\tau)}}{\sqrt{c_0^2 - \dot{\vec{r}}_t(\tau)\cdot\dot{\vec{r}}_t(\tau)}}\right)}_{w} d\tau \qquad (6.9)$$

As in eq. 6.3 a closed form solution for the received signal $g(t)$ is only obtained if the expression for the new variable w can be solved for τ. Otherwise approximations may be found e. g. by linearisation of the argument of the delta function in eq. 6.9 about the time of expansion τ_0.

$$\frac{c_0\tau + \sqrt{\vec{r}_t(\tau)\cdot\vec{r}_t(\tau)}}{\sqrt{c_0^2 - \dot{\vec{r}}_t(\tau)\cdot\dot{\vec{r}}_t(\tau)}} \approx \frac{c_0\tau_0 + \sqrt{\vec{r}_t(\tau_0)\cdot\vec{r}_t(\tau_0)}}{\sqrt{c_0^2 - \dot{\vec{r}}_t(\tau_0)\cdot\dot{\vec{r}}_t(\tau_0)}} + (\tau-\tau_0)\cdot\frac{d}{d\tau}\left.\frac{c_0\tau + \sqrt{\vec{r}_t(\tau)\cdot\vec{r}_t(\tau)}}{\sqrt{c_0^2 - \dot{\vec{r}}_t(\tau)\cdot\dot{\vec{r}}_t(\tau)}}\right|_{\tau=\tau_0}$$

$$\frac{c_0\tau + \sqrt{\vec{r}_t(\tau)\cdot\vec{r}_t(\tau)}}{\sqrt{c_0^2 - \dot{\vec{r}}_t(\tau)\cdot\dot{\vec{r}}_t(\tau)}} \approx \frac{c_0\tau_0 + \sqrt{\vec{r}_t(\tau_0)\cdot\vec{r}_t(\tau_0)}}{\sqrt{c_0^2 - \dot{\vec{r}}_t(\tau_0)\cdot\dot{\vec{r}}_t(\tau_0)}} + (\tau-\tau_0)$$

$$\cdot\left.\frac{\sqrt{c_0^2 - \dot{\vec{r}}_t(\tau)\cdot\dot{\vec{r}}_t(\tau)}\frac{d}{d\tau}\left(c_0\tau + \sqrt{\vec{r}_t(\tau)\cdot\vec{r}_t(\tau)}\right) - \left(c_0\tau + \sqrt{\vec{r}_t(\tau)\cdot\vec{r}_t(\tau)}\right)\frac{d}{d\tau}\sqrt{c_0^2 - \dot{\vec{r}}_t(\tau)\cdot\dot{\vec{r}}_t(\tau)}}{c_0^2 - \dot{\vec{r}}_t(\tau)\cdot\dot{\vec{r}}_t(\tau)}\right|_{\tau=\tau_0}$$

$$\frac{c_0\tau + \sqrt{\vec{r}_t(\tau)\cdot\vec{r}_t(\tau)}}{\sqrt{c_0^2 - \dot{\vec{r}}_t(\tau)\cdot\dot{\vec{r}}_t(\tau)}} \approx \frac{c_0\tau_0 + \sqrt{\vec{r}_t(\tau_0)\cdot\vec{r}_t(\tau_0)}}{\sqrt{c_0^2 - \dot{\vec{r}}_t(\tau_0)\cdot\dot{\vec{r}}_t(\tau_0)}} + (\tau-\tau_0)$$

$$\cdot\frac{\sqrt{c_0^2 - \dot{\vec{r}}_t(\tau_0)\cdot\dot{\vec{r}}_t(\tau_0)}\left(c_0 + \frac{\vec{r}_t(\tau_0)\cdot\dot{\vec{r}}_t(\tau_0)}{\sqrt{\vec{r}_t(\tau_0)\cdot\vec{r}_t(\tau_0)}}\right) - \left(c_0\tau_0 + \sqrt{\vec{r}_t(\tau_0)\cdot\vec{r}_t(\tau_0)}\right)\frac{\dot{\vec{r}}_t(\tau_0)\cdot\ddot{\vec{r}}_t(\tau_0)}{\sqrt{c_0^2 - \dot{\vec{r}}_t(\tau_0)\cdot\dot{\vec{r}}_t(\tau_0)}}}{c_0^2 - \dot{\vec{r}}_t(\tau_0)\cdot\dot{\vec{r}}_t(\tau_0)}$$

Using this expression in eq. 6.9 the expression for w in the argument of the delta function can be solved for τ.

$$\tau \approx \tau_0 + \frac{\sqrt{c_0^2 - \vec{\dot{r}}_t(\tau_0) \cdot \vec{\dot{r}}_t(\tau_0)} \left(t - w - \frac{c_0 \tau_0 + \sqrt{\vec{r}_t(\tau_0) \cdot \vec{r}_t(\tau_0)}}{\sqrt{c_0^2 - \vec{\dot{r}}_t(\tau_0) \cdot \vec{\dot{r}}_t(\tau_0)}} \right)}{c_0 + \frac{\vec{r}_t(\tau_0) \cdot \vec{\dot{r}}_t(\tau_0)}{\sqrt{\vec{r}_t(\tau_0) \cdot \vec{r}_t(\tau_0)}} - \left(c_0 \tau_0 + \sqrt{\vec{r}_t(\tau_0) \cdot \vec{r}_t(\tau_0)} \right) \frac{\vec{\dot{r}}_t(\tau_0) \cdot \vec{\ddot{r}}_t(\tau_0)}{c_0^2 - \vec{\dot{r}}_t(\tau_0) \cdot \vec{\dot{r}}_t(\tau_0)}} \quad (6.10a)$$

$$\frac{d\tau}{dw} \approx -\frac{\sqrt{c_0^2 - \vec{\dot{r}}_t(\tau_0) \cdot \vec{\dot{r}}_t(\tau_0)}}{c_0 + \frac{\vec{r}_t(\tau_0) \cdot \vec{\dot{r}}_t(\tau_0)}{\sqrt{\vec{r}_t(\tau_0) \cdot \vec{r}_t(\tau_0)}} - \left(c_0 \tau_0 + \sqrt{\vec{r}_t(\tau_0) \cdot \vec{r}_t(\tau_0)} \right) \frac{\vec{\dot{r}}_t(\tau_0) \cdot \vec{\ddot{r}}_t(\tau_0)}{c_0^2 - \vec{\dot{r}}_t(\tau_0) \cdot \vec{\dot{r}}_t(\tau_0)}} \quad (6.10b)$$

If a sinusoidal waveform is used as transmitted signal the instantaneous frequency of the received signal is calculated according to eq. 6.7.

$$\frac{f(\tau_0)}{f_0} = \frac{\sqrt{c_0^2 - \vec{\dot{r}}_t(\tau_0) \cdot \vec{\dot{r}}_t(\tau_0)}}{c_0 + \frac{\vec{r}_t(\tau_0) \cdot \vec{\dot{r}}_t(\tau_0)}{\sqrt{\vec{r}_t(\tau_0) \cdot \vec{r}_t(\tau_0)}} - \left(c_0 \tau_0 + \sqrt{\vec{r}_t(\tau_0) \cdot \vec{r}_t(\tau_0)} \right) \frac{\vec{\dot{r}}_t(\tau_0) \cdot \vec{\ddot{r}}_t(\tau_0)}{c_0^2 - \vec{\dot{r}}_t(\tau_0) \cdot \vec{\dot{r}}_t(\tau_0)}} \quad (6.11)$$

6.2 Stationary Transmitter and non-uniform Motion of Receiver

Now a stationary transmitter located at the point of origin is considered. The receiver moves along a trajectory $\vec{r}_r(t) = (x_r(t), y_r(t), z_r(t))$. At time τ the transmitter emits a spherical wave. Eq. 6.12 characterises the evolution of the sphere as a function of time.

$$\vec{r} \cdot \vec{r} = c_0^2 (t - \tau)^2 \quad (6.12)$$

The squared radius of the sphere at time t is on the right-hand side of the equation. The radius is proportional to the velocity of propagation c_0 of the signal. At time t_r the spherical wave arrives at the receiver. This time can be calculated using the trajectory of the receiver in eq. 6.12.

$$\vec{r}_r(t_r) \cdot \vec{r}_r(t_r) = c_0^2 (t_r - \tau)^2 \Rightarrow \vec{r}_r(t_r) \cdot \vec{r}_r(t_r) = c_0^2 t_r^2 - 2c_0^2 \tau t_r + c_0^2 \tau^2$$

6.2.1 Non-relativistic Doppler Effect

Solving this equation for τ enables the direct calculation of the received signal g(t) according to eq. 4.6a. The receiving time t_r is replaced by t.

$$\tau^2 - 2t\tau + \frac{c_0^2 t^2 - \vec{r}_r(t) \cdot \vec{r}_r(t)}{c_0^2} = 0 \Rightarrow \tau = t - c_0^{-1}\sqrt{\vec{r}_r(t) \cdot \vec{r}_r(t)} \qquad (6.13a)$$

$$\frac{d\tau}{dt} = 1 - c_0^{-1}\frac{d}{dt}\sqrt{\vec{r}_r(t) \cdot \vec{r}_r(t)} = 1 - c_0^{-1}\frac{\vec{r}_r(t) \cdot \dot{\vec{r}}_r(t)}{\sqrt{\vec{r}_r(t) \cdot \vec{r}_r(t)}} \qquad (6.13b)$$

$$g(t) = \left| 1 - c_0^{-1}\frac{\vec{r}_r(t) \cdot \dot{\vec{r}}_r(t)}{\sqrt{\vec{r}_r(t) \cdot \vec{r}_r(t)}} \right| \cdot s\left(t - c_0^{-1}\sqrt{\vec{r}_r(t) \cdot \vec{r}_r(t)}\right) \qquad (6.13c)$$

The minus sign in front of the square root is selected because of the causality condition $\tau \leq t_r$.

In the special case $\vec{r}_r(t) = (v_r t, y_0, 0)$ eq. 6.13c is identical to eq. 4.10c. A closed form solution for the time variant impulse response h(t,τ), according to eq. 2.2, is only obtained if eq. 6.13a can be solved for t. Otherwise approximations may be found e. g. by linearisation of the square root in eq. 6.13a about the time of expansion t_0.

6.2.2 Relativistic Doppler Effect

The Lorentz transform is applied to the transmission time τ of the signal according to eq. 6.13a using the instantaneous velocity $\dot{\vec{r}}_r(t)$ of the receiver at time $t = t_r$.

$$\tau' = \frac{1}{\sqrt{1 - v_r^2(t)/c_0^2}}\left(t - c_0^{-1}\sqrt{\vec{r}_r(t) \cdot \vec{r}_r(t)}\right) = \frac{c_0 t - \sqrt{\vec{r}_r(t) \cdot \vec{r}_r(t)}}{\sqrt{c_0^2 - \dot{\vec{r}}_r(t) \cdot \dot{\vec{r}}_r(t)}} \qquad (6.14a)$$

The quotient rule for derivatives is applied to eq. 6.14a.

$$\frac{d\tau'}{dt} = \frac{\sqrt{c_0^2 - \dot{\vec{r}}_r(t) \cdot \dot{\vec{r}}_r(t)}\frac{d}{dt}\left(c_0 t - \sqrt{\vec{r}_r(t) \cdot \vec{r}_r(t)}\right) - \left(c_0 t - \sqrt{\vec{r}_r(t) \cdot \vec{r}_r(t)}\right)\frac{d}{dt}\sqrt{c_0^2 - \dot{\vec{r}}_r(t) \cdot \dot{\vec{r}}_r(t)}}{c_0^2 - \dot{\vec{r}}_r(t) \cdot \dot{\vec{r}}_r(t)}$$

$$\frac{d\tau'}{dt} = \frac{\sqrt{c_0^2 - \dot{\vec{r}}_r(t) \cdot \dot{\vec{r}}_r(t)}\left(c_0 - \frac{\vec{r}_r(t) \cdot \dot{\vec{r}}_r(t)}{\sqrt{\vec{r}_r(t) \cdot \vec{r}_r(t)}}\right) + \left(c_0 t - \sqrt{\vec{r}_r(t) \cdot \vec{r}_r(t)}\right)\frac{\dot{\vec{r}}_r(t) \cdot \ddot{\vec{r}}_r(t)}{\sqrt{c_0^2 - \dot{\vec{r}}_r(t) \cdot \dot{\vec{r}}_r(t)}}}{c_0^2 - \dot{\vec{r}}_r(t) \cdot \dot{\vec{r}}_r(t)} \qquad (6.14b)$$

Combining eqs. 6.14a and 6.14b the received signal g(t) is obtained.

$$g(t) = \left| \frac{c_0 - \frac{\vec{r}_r(t) \cdot \dot{\vec{r}}_r(t)}{\sqrt{\vec{r}_r(t) \cdot \vec{r}_r(t)}}}{\sqrt{c_0^2 - \dot{\vec{r}}_r(t) \cdot \dot{\vec{r}}_r(t)}} + \frac{\left(c_0 t - \sqrt{\vec{r}_r(t) \cdot \vec{r}_r(t)}\right) \dot{\vec{r}}_r(t) \cdot \ddot{\vec{r}}_r(t)}{\sqrt{c_0^2 - \dot{\vec{r}}_r(t) \cdot \dot{\vec{r}}_r(t)}^3} \right| \cdot s\left(\frac{c_0 t - \sqrt{\vec{r}_r(t) \cdot \vec{r}_r(t)}}{\sqrt{c_0^2 - \dot{\vec{r}}_r(t) \cdot \dot{\vec{r}}_r(t)}} \right) \quad (6.14c)$$

In the special case $\vec{r}_r(t) = (v_r t, y_0, 0)$ eq. 6.14c is identical to eq. 4.28.

6.3 Experimental Test

Fig. 25 shows an experimental setup consisting of a loudspeaker (acoustic transmitter) in circular motion and a stationary microphone (acoustic receiver). This setup is used for experiments on spectral analysis.

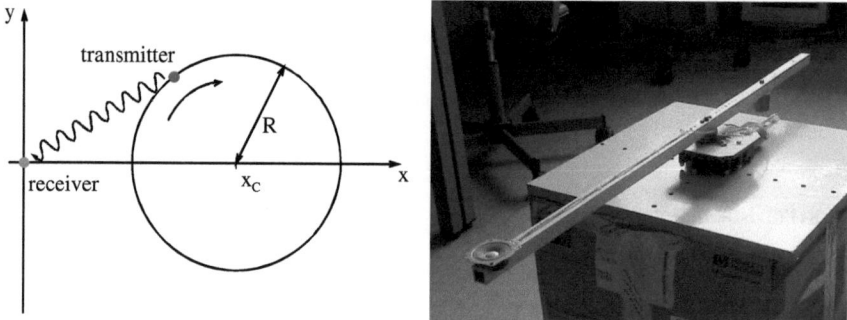

Fig. 25: Stationary receiver and transmitter in circular motion

The trajectory $\vec{r}_r(t)$ resp. the instantaneous velocity $\dot{\vec{r}}_r(t)$ of the transmitter are:

$$\vec{r}_r(t) = \left(R\cos(2\pi f_R t) + x_C, R\sin(2\pi f_R t), 0\right)$$
$$\dot{\vec{r}}_r(t) = \left(-2\pi f_R R\sin(2\pi f_R t), 2\pi f_R R\cos(2\pi f_R t), 0\right)$$

Using these expressions in eq. 6.7 the normalised instantaneous frequency is obtained.

$$\frac{f(\tau_0)}{f_0} \approx \frac{c_0 \sqrt{\left(R\cos(2\pi f_R \tau_0) + x_C\right)^2 + R^2 \sin^2(2\pi f_R \tau_0)}}{c_0 \sqrt{\left(R\cos(2\pi f_R \tau_0) + x_C\right)^2 + R^2 \sin^2(2\pi f_R \tau_0)} - 2\pi f_R R x_C \sin(2\pi f_R \tau_0)}$$

Using trigonometric identities this equation can be simplified.

$$\frac{f(\tau_0)}{f_0} \approx \frac{c_0\sqrt{R^2 + 2Rx_C \cos(2\pi f_R \tau_0) + x_C^2}}{c_0\sqrt{R^2 + 2Rx_C \cos(2\pi f_R \tau_0) + x_C^2} - 2\pi f_R Rx_C \sin(2\pi f_R \tau_0)} \quad (6.15)$$

Using the trajectory $\vec{r}_t(t)$ and the instantaneous velocity $\dot{\vec{r}}_t(t)$ in the relativistic equation 6.7 the normalised instantaneous frequency could be calculated for relativistic motion. Of course this calculation is not applicable to the acoustic experiment.

The bright curve in fig. 26 shows the normalised instantaneous frequency which was calculated using eq. 6.15. This curve was plotted over a measured spectrogram. The parameters used for the experiment are: f_0 = 4 kHz, f_R = 1,8 Hz, R = 0,52 m, c_0 = 345 m/s, x_C = 0,57 m. The match of theory and experiment is remarkably good.

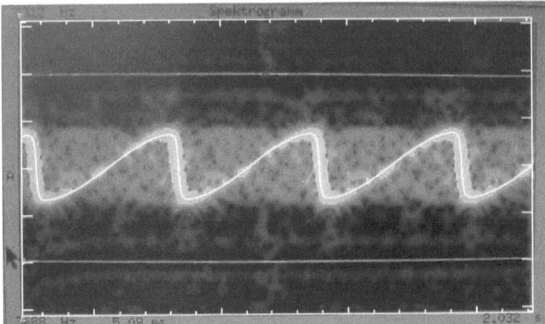

Fig. 26: Comparison of measured and calculated instantaneous frequencies

7 Conclusions

From the derivation of a mathematical description in terms of system theory we gained profound insight into the generation of the Doppler Effect. Basically it is caused by time-variant signal delay. Three effects determine the received signal: time scaling, amplitude scaling and time delay. Each of these effects can be a function of time.

The transfer of the results from the non-relativistic to the relativistic Doppler Effect is accomplished using a simple purely temporal Lorentz transform. The validity of this approach is confirmed by application to various special cases and comparison against results from Einstein's paper on special relativity including the Doppler Effect for reflected signals e. g. in the case of monostatic radar.

For transmitters and/or receivers in arbitrary uniform motion the Doppler Effect can be derived for arbitrary waveforms of the transmitted signal.

For a stationary transmitter and a receiver in arbitrary non-uniform motion the received signal can also be calculated analytically for arbitrary waveforms of the transmitted signal. For a stationary receiver and a transmitter in arbitrary non-uniform motion the instantaneous frequency of the received signal can currently be calculated analytically for sinusoidal transmitted signals.

However there are some open problems and questions.

Is it possible to transfer the results for monostatic radar to bistatic resp. multistatic radar using a reasonable amount of mathematics?

Is it possible to incorporate general relativity for non-uniform motion?

Additional signal attenuation depending on the distance of transmitter and receiver as well as on the spatial orientation of transmitter and receiver in space should be included.

Especially the propagation of acoustic signals may be influenced by dispersion, in other words the velocity of propagation of the signal may be frequency dependent.

The calculation of Auto Correlation Functions (ACF) showed that currently the application of the approach from system theory to random signals is somewhat limited.

Radar – reflector configurations in non-uniform motion have not yet been considered.

A   Appendices

A.1   Derivation of Equation 4.24

$$f(x) = f_0 \frac{c_0}{\sqrt{c_0^2 - v_t^2}} \left(1 - \frac{v_t}{c_0} \frac{v_t \frac{c_0}{\sqrt{c_0^2 - v_t^2}} \left(\frac{x}{v_t} + \frac{1}{c_0} \sqrt{v_t^2 \frac{x^2}{v_t^2} + y_0^2} \right)}{\sqrt{v_t^2 \left(\frac{c_0}{\sqrt{c_0^2 - v_t^2}} \left(\frac{x}{v_t} + \frac{1}{c_0} \sqrt{v_t^2 \frac{x^2}{v_t^2} + y_0^2} \right) \right)^2 + y_0^2}} \right)$$

$$f(x) = f_0 \frac{c_0}{\sqrt{c_0^2 - v_t^2}} \left(1 - \frac{v_t}{c_0} \frac{x + \frac{v_t}{c_0} \sqrt{x^2 + y_0^2}}{\sqrt{\left(x + \frac{v_t}{c_0} \sqrt{x^2 + y_0^2} \right)^2 + \frac{c_0^2 - v_t^2}{c_0^2} y_0^2}} \right)$$

$$f(x) = f_0 \frac{c_0}{\sqrt{c_0^2 - v_t^2}} \left(1 - \frac{v_t}{c_0} \frac{x + \frac{v_t}{c_0} \sqrt{x^2 + y_0^2}}{\sqrt{x^2 + 2x \frac{v_t}{c_0} \sqrt{x^2 + y_0^2} + \frac{v_t^2}{c_0^2} \left(x^2 + \cancel{y_0^2} \right) + \frac{c_0^2 - \cancel{v_t^2}}{c_0^2} y_0^2}} \right)$$

$$f(x) = f_0 \frac{c_0}{\sqrt{c_0^2 - v_t^2}} \left(1 - \frac{v_t}{c_0} \frac{x + \frac{v_t}{c_0} \sqrt{x^2 + y_0^2}}{\sqrt{\left(\frac{v_t}{c_0} x + \sqrt{x^2 + y_0^2} \right)^2}} \right) = f_0 \frac{c_0}{\sqrt{c_0^2 - v_t^2}} \left(1 - \frac{v_t}{c_0} \frac{x + \frac{v_t}{c_0} \sqrt{x^2 + y_0^2}}{\frac{v_t}{c_0} x + \sqrt{x^2 + y_0^2}} \right)$$

$$f(x) = f_0 \frac{c_0}{\sqrt{c_0^2 - v_t^2}} \left(\frac{\cancel{\frac{v_t}{c_0}} x + \sqrt{x^2 + y_0^2}}{\frac{v_t}{c_0} x + \sqrt{x^2 + y_0^2}} - \frac{v_t}{c_0} \frac{\cancel{x} + \frac{v_t}{c_0} \sqrt{x^2 + y_0^2}}{\frac{v_t}{c_0} x + \sqrt{x^2 + y_0^2}} \right) = f_0 \frac{\sqrt{1 - v_t^2/c_0^2}}{1 + \frac{v_t}{c_0} \frac{x}{\sqrt{x^2 + y_0^2}}}$$

A.2 Relativistic Subtraction of Velocity Vectors

The calculation is done using an approach which can be found in textbooks on special relativity, such as [4, 6]. The velocity vector $\vec{v}_t$ of the transmitter is decomposed into the vector $\vec{v}_{t\|}$ parallel to the velocity vector $\vec{v}_r$ of the receiver and the vector $\vec{v}_{t\perp}$ perpendicular to the velocity vector of the receiver.

$$\vec{v}_t = \vec{v}_{t\|} + \vec{v}_{t\perp}$$

The orthogonal vector components can be calculated using the scalar product.

$$\vec{v}_{t\|} = \left(\vec{v}_t \cdot \frac{\vec{v}_r}{\sqrt{\vec{v}_r \cdot \vec{v}_r}}\right) \frac{\vec{v}_r}{\sqrt{\vec{v}_r \cdot \vec{v}_r}} = \frac{\vec{v}_t \cdot \vec{v}_r}{\vec{v}_r \cdot \vec{v}_r} \vec{v}_r \qquad \vec{v}_{t\perp} = \vec{v}_t - \vec{v}_{t\|} = \vec{v}_t - \frac{\vec{v}_t \cdot \vec{v}_r}{\vec{v}_r \cdot \vec{v}_r} \vec{v}_r$$

The relativistic subtraction of velocity vectors $\vec{v}_d$ is also decomposed into a vector $\vec{v}_{d\|}$ parallel to the velocity vector of the receiver and a vector $\vec{v}_{d\perp}$ perpendicular to the velocity vector of the receiver. $\vec{v}_{d\|}$ resp. $\vec{v}_{d\perp}$ are calculated using the approach from [4, 6].

$$\vec{v}_{d\|} = \frac{\vec{v}_{t\|} - \vec{v}_r}{1 - \vec{v}_{t\|} \cdot \vec{v}_r / c_0^2} = \frac{\vec{v}_t \cdot \vec{v}_r / \vec{v}_r \cdot \vec{v}_r - 1}{1 - \vec{v}_{t\|} \cdot \vec{v}_r / c_0^2} \vec{v}_r$$

$$\vec{v}_{d\perp} = \vec{v}_{t\perp} \frac{\sqrt{1 - \vec{v}_r \cdot \vec{v}_r / c_0^2}}{1 - \vec{v}_{t\|} \cdot \vec{v}_r / c_0^2} = \left(\vec{v}_t - \frac{\vec{v}_t \cdot \vec{v}_r}{\vec{v}_r \cdot \vec{v}_r} \vec{v}_r\right) \frac{\sqrt{1 - \vec{v}_r \cdot \vec{v}_r / c_0^2}}{1 - \vec{v}_{t\|} \cdot \vec{v}_r / c_0^2}$$

Using $\vec{v}_{t\|} = \vec{v}_t - \vec{v}_{t\perp}$ and $\vec{v}_{t\|} \cdot \vec{v}_r = \vec{v}_t \cdot \vec{v}_r - \underbrace{\vec{v}_{t\perp} \cdot \vec{v}_r}_{0} = \vec{v}_t \cdot \vec{v}_r$ the expression for $\vec{v}_d$ can be simplified.

$$\vec{v}_d = \vec{v}_{d\|} + \vec{v}_{d\perp} = \frac{\vec{v}_t \cdot \vec{v}_r / \vec{v}_r \cdot \vec{v}_r - 1}{1 - \vec{v}_t \cdot \vec{v}_r / c_0^2} \vec{v}_r + \left(\vec{v}_t - \frac{\vec{v}_t \cdot \vec{v}_r}{\vec{v}_r \cdot \vec{v}_r} \vec{v}_r\right) \frac{\sqrt{1 - \vec{v}_r \cdot \vec{v}_r / c_0^2}}{1 - \vec{v}_t \cdot \vec{v}_r / c_0^2}$$

In the special case $\vec{v}_t \| \vec{v}_r$, i.e. $\vec{v}_r = a\vec{v}_t$, an expression similar to eq. 3.20 can be derived.

$$\vec{v}_t \| \vec{v}_r \Rightarrow \vec{v}_d = \frac{\vec{v}_t \cdot a\vec{v}_t / a\vec{v}_t \cdot a\vec{v}_t - 1}{1 - \vec{v}_t \cdot a\vec{v}_t / c_0^2} a\vec{v}_t + \underbrace{\left(\vec{v}_t - \frac{\vec{v}_t \cdot a\vec{v}_t}{a\vec{v}_t \cdot a\vec{v}_t} a\vec{v}_t\right)}_{0} \frac{\sqrt{1 - a\vec{v}_t \cdot a\vec{v}_t / c_0^2}}{1 - \vec{v}_t \cdot a\vec{v}_t / c_0^2}$$

$$\vec{v}_d = \frac{\vec{v}_t - a\vec{v}_t}{1 - \vec{v}_t \cdot a\vec{v}_t / c_0^2} = \frac{\vec{v}_t - \vec{v}_r}{1 - \vec{v}_t \cdot \vec{v}_r / c_0^2}$$

Two further special cases are: $\quad \vec{v}_r = \vec{0} \Rightarrow \vec{v}_d = \vec{v}_t \quad$ resp. $\quad \vec{v}_t = \vec{0} \Rightarrow \vec{v}_d = -\vec{v}_r$

Of course the relativistic subtraction of velocities is constrained to $-c_0 \leq v_d \leq +c_0$.

a) If the receiver is moving in the direction of an arbitrary unit vector $\vec{e}$ ($|\vec{e}|=1$) with the velocity of light c_0 the relativistic subtraction of velocity vectors is equal to the reversed velocity vector of the receiver.

$$\vec{v}_r = c_0\vec{e} \Rightarrow \vec{v}_d = \underbrace{\frac{\vec{v}_t \cdot c_0\vec{e}/c_0\vec{e} \cdot c_0\vec{e} - 1}{1-\vec{v}_t \cdot c_0\vec{e}/c_0^2}}_{-1}c_0\vec{e} + \left(\vec{v}_t - \frac{\vec{v}_t \cdot c_0\vec{e}}{c_0\vec{e} \cdot c_0\vec{e}}c_0\vec{e}\right)\underbrace{\frac{\sqrt{1-c_0\vec{e} \cdot c_0\vec{e}/c_0^2}}{1-\vec{v}_t \cdot c_0\vec{e}/c_0^2}}_{0} = -c_0\vec{e}$$

b) If the transmitter is moving in the direction of an arbitrary unit vector $\vec{e}$ ($|\vec{e}|=1$) with the velocity of light c_0 the modulus of the relativistic subtraction of velocity vectors is equal to the modulus of the velocity vector of the transmitter. In this case the direction of the relativistic subtraction of velocity vectors differs from the direction of the velocity vector of the transmitter.

Using the orthogonality relation $\vec{v}_d \cdot \vec{v}_d = \vec{v}_{d\parallel} \cdot \vec{v}_{d\parallel} + \vec{v}_{d\perp} \cdot \vec{v}_{d\perp}$ the squared modulus of the relativistic subtraction of velocity vectors is obtained.

$$\vec{v}_d \cdot \vec{v}_d = \left(\frac{\vec{v}_t \cdot \vec{v}_r - \vec{v}_r \cdot \vec{v}_r}{1-\vec{v}_t \cdot \vec{v}_r/c_0^2}\right)^2 \frac{\vec{v}_r}{\vec{v}_r \cdot \vec{v}_r} \cdot \frac{\vec{v}_r}{\vec{v}_r \cdot \vec{v}_r} + \left(\vec{v}_t - \frac{\vec{v}_t \cdot \vec{v}_r}{\vec{v}_r \cdot \vec{v}_r}\vec{v}_r\right) \cdot \left(\vec{v}_t - \frac{\vec{v}_t \cdot \vec{v}_r}{\vec{v}_r \cdot \vec{v}_r}\vec{v}_r\right)\frac{1-\vec{v}_t \cdot \vec{v}_r/c_0^2}{\left(1-\vec{v}_t \cdot \vec{v}_r/c_0^2\right)^2}$$

$$\vec{v}_d \cdot \vec{v}_d = \frac{1}{\left(1-\vec{v}_t \cdot \vec{v}_r/c_0^2\right)^2}\left((\vec{v}_t \cdot \vec{v}_r - \vec{v}_r \cdot \vec{v}_r)^2\frac{1}{\vec{v}_r \cdot \vec{v}_r} + \left(\vec{v}_t \cdot \vec{v}_t - \frac{(\vec{v}_t \cdot \vec{v}_r)^2}{\vec{v}_r \cdot \vec{v}_r}\right)(1-\vec{v}_t \cdot \vec{v}_r/c_0^2)\right)$$

$$\vec{v}_d \cdot \vec{v}_d = \frac{(\vec{v}_t \cdot \vec{v}_r - \vec{v}_r \cdot \vec{v}_r)^2 + (\vec{v}_t \cdot \vec{v}_t \vec{v}_r \cdot \vec{v}_r - (\vec{v}_t \cdot \vec{v}_r)^2)(1-\vec{v}_t \cdot \vec{v}_r/c_0^2)}{\left(1-\vec{v}_t \cdot \vec{v}_r/c_0^2\right)^2 \vec{v}_r \cdot \vec{v}_r}$$

$$\vec{v}_d \cdot \vec{v}_d = \frac{\cancel{(\vec{v}_t \cdot \vec{v}_r)^2} - 2\vec{v}_t \cdot \vec{v}_r \vec{v}_r \cdot \vec{v}_r + (\vec{v}_r \cdot \vec{v}_r)^2 + \vec{v}_t \cdot \vec{v}_t \vec{v}_r \cdot \vec{v}_r - \cancel{(\vec{v}_t \cdot \vec{v}_r)^2} - \vec{v}_t \cdot \vec{v}_t \vec{v}_r \cdot \vec{v}_r \frac{\vec{v}_t \cdot \vec{v}_r}{c_0^2} + (\vec{v}_t \cdot \vec{v}_r)^2 \frac{\vec{v}_t \cdot \vec{v}_r}{c_0^2}}{\left(1-\vec{v}_t \cdot \vec{v}_r/c_0^2\right)^2 \vec{v}_r \cdot \vec{v}_r}$$

$$\vec{v}_d \cdot \vec{v}_d = \frac{1}{\left(1-\vec{v}_t \cdot \vec{v}_r/c_0^2\right)^2}\frac{1}{\vec{v}_r \cdot \vec{v}_r}\left(c_0^2 \vec{v}_r \cdot \vec{v}_r + (\vec{v}_t \cdot \vec{v}_r)^2\frac{\vec{v}_r \cdot \vec{v}_r}{c_0^2} - 2\vec{v}_t \cdot \vec{v}_r \vec{v}_r \cdot \vec{v}_r\right)$$

$$\vec{v}_d \cdot \vec{v}_d = \frac{1}{\left(1-\vec{v}_t \cdot \vec{v}_r/c_0^2\right)^2}\left(c_0^2 - 2\vec{v}_t \cdot \vec{v}_r + \frac{(\vec{v}_t \cdot \vec{v}_r)^2}{c_0^2}\right) = \frac{c_0^2}{\left(1-\vec{v}_t \cdot \vec{v}_r/c_0^2\right)^2}\left(1 - 2\frac{\vec{v}_t \cdot \vec{v}_r}{c_0^2} + \left(\frac{\vec{v}_t \cdot \vec{v}_r}{c_0^2}\right)^2\right)$$

$$\vec{v}_d \cdot \vec{v}_d = \frac{c_0^2}{\left(1-\vec{v}_t \cdot \vec{v}_r/c_0^2\right)^2}\left(1-\frac{\vec{v}_t \cdot \vec{v}_r}{c_0^2}\right)^2 = c_0^2$$

If the velocity vector $\vec{v}_r$ of the receiver is decomposed into a vector $\vec{v}_{r\|}$ parallel to the velocity vector $\vec{v}_t$ of the transmitter and a vector $\vec{v}_{r\perp}$ perpendicular to the velocity vector of the transmitter the properties of case a) - determination of vector - and b) - determination of modulus only - are interchanged.

The difference results from the fact that in case a) the coordinate system of the receiver is used while in case b) the coordinate system of the transmitter is used.

A.3 Derivation of Equation 5.9

$$\tau + \frac{1}{c_0}\sqrt{v^2\tau^2 + \left(1-\frac{v^2}{c_0^2}\right)y_0^2} = t_r - \frac{1}{c_0}\sqrt{v^2t_r^2 + \left(1-\frac{v^2}{c_0^2}\right)y_0^2}$$

$$\sqrt{v^2t_r^2 + \left(1-\frac{v^2}{c_0^2}\right)y_0^2} = c_0 t_r - c_0\tau - \sqrt{v^2\tau^2 + \left(1-\frac{v^2}{c_0^2}\right)y_0^2}$$

$$v^2 t_r^2 + \left(1-\frac{v^2}{c_0^2}\right)y_0^2 = \left(c_0 t_r - \left(c_0\tau + \sqrt{v^2\tau^2 + \left(1-\frac{v^2}{c_0^2}\right)y_0^2}\right)\right)^2$$

$$v^2 t_r^2 + \left(1-\frac{v^2}{c_0^2}\right)y_0^2 = c_0^2 t_r^2 - 2\left(c_0\tau + \sqrt{v^2\tau^2 + \left(1-\frac{v^2}{c_0^2}\right)y_0^2}\right)c_0 t_r + \left(c_0\tau + \sqrt{v^2\tau^2 + \left(1-\frac{v^2}{c_0^2}\right)y_0^2}\right)^2$$

$$t_r^2 - 2\frac{c_0\tau + \sqrt{v^2\tau^2 + \left(1-\frac{v^2}{c_0^2}\right)y_0^2}}{c_0^2 - v^2} c_0 t_r + \frac{\left(c_0\tau + \sqrt{v^2\tau^2 + \left(1-\frac{v^2}{c_0^2}\right)y_0^2}\right)^2 - \left(1-\frac{v^2}{c_0^2}\right)y_0^2}{c_0^2 - v^2} = 0$$

$$t_r = \frac{c_0\tau + \sqrt{v^2\tau^2 + \left(1-\frac{v^2}{c_0^2}\right)y_0^2}}{c_0^2 - v^2} c_0$$
$$+ \sqrt{\left(\frac{c_0\tau + \sqrt{v^2\tau^2 + \left(1-\frac{v^2}{c_0^2}\right)y_0^2}}{c_0^2 - v^2} c_0\right)^2 - \frac{\left(c_0\tau + \sqrt{v^2\tau^2 + \left(1-\frac{v^2}{c_0^2}\right)y_0^2}\right)^2 - \left(1-\frac{v^2}{c_0^2}\right)y_0^2}{c_0^2 - v^2}}$$

$$t_r = \frac{c_0^2}{c_0^2 - v^2}\left(\tau + \sqrt{\frac{v^2}{c_0^2}\tau^2 + \frac{y_0^2}{c_0^2}\frac{c_0^2 - v^2}{c_0^2}}\right)$$
$$+ \sqrt{\left(\frac{c_0^2}{c_0^2 - v^2}\left(\tau + \sqrt{\frac{v^2}{c_0^2}\tau^2 + \frac{y_0^2}{c_0^2}\frac{c_0^2 - v^2}{c_0^2}}\right)\right)^2 - \frac{\cancel{c_0^2} - v^2}{c_0^2 - v^2}\frac{c_0^2}{c_0^2 - v^2}\left(\tau + \sqrt{\frac{v^2}{c_0^2}\tau^2 + \frac{y_0^2}{c_0^2}\frac{c_0^2 - v^2}{c_0^2}}\right)^2 + \frac{y_0^2}{c_0^2}}$$

$$t_r = \frac{c_0^2}{c_0^2 - v^2}\left(\tau + \sqrt{\frac{v^2}{c_0^2}\tau^2 + \frac{y_0^2}{c_0^2}\frac{c_0^2 - v^2}{c_0^2}}\right) + \sqrt{\frac{c_0^2 v^2}{(c_0^2 - v^2)^2}\left(\tau + \sqrt{\frac{v^2}{c_0^2}\tau^2 + \frac{y_0^2}{c_0^2}\frac{c_0^2 - v^2}{c_0^2}}\right)^2 + \frac{y_0^2}{c_0^2}}$$

A.4 Derivation of Equation 5.15a

$$\tau'\sqrt{1-\frac{v^2}{c_0^2}} = \frac{c_0^2}{c_0^2-v^2}\left(t'\sqrt{1-\frac{v^2}{c_0^2}} - \sqrt{\frac{v^2}{c_0^2}\left(t'\sqrt{1-\frac{v^2}{c_0^2}}\right)^2 + \frac{y_0^2}{c_0^2}\frac{c_0^2-v^2}{c_0^2}}\right)$$

$$-\sqrt{\frac{c_0^2 v^2}{\left(c_0^2-v^2\right)^2}\left(t'\sqrt{1-\frac{v^2}{c_0^2}} - \sqrt{\frac{v^2}{c_0^2}\left(t'\sqrt{1-\frac{v^2}{c_0^2}}\right)^2 + \frac{y_0^2}{c_0^2}\frac{c_0^2-v^2}{c_0^2}}\right)^2 + \frac{y_0^2}{c_0^2}}$$

$$\tau'\sqrt{\frac{c_0^2-v^2}{c_0^2}} = \frac{c_0^2}{c_0^2-v^2}\left(t'\sqrt{\frac{c_0^2-v^2}{c_0^2}} - \sqrt{\frac{v^2}{c_0^2}t'^2\frac{c_0^2-v^2}{c_0^2} + \frac{y_0^2}{c_0^2}\frac{c_0^2-v^2}{c_0^2}}\right)$$

$$-\sqrt{\frac{c_0^2 v^2}{\left(c_0^2-v^2\right)^2}\left(t'\sqrt{\frac{c_0^2-v^2}{c_0^2}} - \sqrt{\frac{v^2}{c_0^2}t'^2\frac{c_0^2-v^2}{c_0^2} + \frac{y_0^2}{c_0^2}\frac{c_0^2-v^2}{c_0^2}}\right)^2 + \frac{y_0^2}{c_0^2}}$$

$$\tau'\sqrt{\frac{c_0^2-v^2}{c_0^2}} =$$

$$\frac{c_0^2}{c_0^2-v^2}\sqrt{\frac{c_0^2-v^2}{c_0^2}}\left(t' - \sqrt{\frac{v^2}{c_0^2}t'^2 + \frac{y_0^2}{c_0^2}}\right) - \sqrt{\frac{c_0^2 v^2}{\left(c_0^2-v^2\right)^2}\left(t'\sqrt{\frac{c_0^2-v^2}{c_0^2}} - \sqrt{\frac{c_0^2-v^2}{c_0^2}}\sqrt{\frac{v^2}{c_0^2}t'^2 + \frac{y_0^2}{c_0^2}}\right)^2 + \frac{y_0^2}{c_0^2}}$$

$$\tau' = \frac{c_0^2}{c_0^2-v^2}\left(t' - \sqrt{\frac{v^2}{c_0^2}t'^2 + \frac{y_0^2}{c_0^2}}\right) - \frac{1}{\sqrt{\frac{c_0^2-v^2}{c_0^2}}}\sqrt{\frac{v^2}{c_0^2-v^2}\left(t' - \sqrt{\frac{v^2}{c_0^2}t'^2 + \frac{y_0^2}{c_0^2}}\right)^2 + \frac{y_0^2}{c_0^2}}$$

$$\tau' = \frac{c_0^2}{c_0^2-v^2}\left(t' - \sqrt{\frac{v^2}{c_0^2}t'^2 + \frac{y_0^2}{c_0^2}}\right) - \frac{c_0}{c_0^2-v^2}\sqrt{v^2\left(t' - \sqrt{\frac{v^2}{c_0^2}t'^2 + \frac{y_0^2}{c_0^2}}\right)^2 + \left(c_0^2-v^2\right)\frac{y_0^2}{c_0^2}}$$

$$\tau' = \frac{c_0^2}{c_0^2-v^2}\left(t' - \sqrt{\frac{v^2}{c_0^2}t'^2 + \frac{y_0^2}{c_0^2}} - \sqrt{\frac{v^2}{c_0^2}\left(t' - \sqrt{\frac{v^2}{c_0^2}t'^2 + \frac{y_0^2}{c_0^2}}\right)^2 + \frac{c_0^2-v^2}{c_0^2}\frac{y_0^2}{c_0^2}}\right)$$

$$\tau' = \frac{c_0^2}{c_0^2-v^2}\left(t' - \sqrt{\frac{v^2}{c_0^2}t'^2 + \frac{y_0^2}{c_0^2}} - \sqrt{\frac{v^2}{c_0^2}t'^2 - 2\frac{v^2}{c_0^2}t'\sqrt{\frac{v^2}{c_0^2}t'^2 + \frac{y_0^2}{c_0^2}} + \frac{v^2}{c_0^2}\left(\frac{v^2}{c_0^2}t'^2 + \cancel{\frac{y_0^2}{c_0^2}}\right) + \frac{c_0^2\cancel{-v^2}}{c_0^2}\frac{y_0^2}{c_0^2}}\right)$$

$$\tau' = \frac{c_0^2}{c_0^2 - v^2}\left(t' - \sqrt{\frac{v^2}{c_0^2}t'^2 + \frac{y_0^2}{c_0^2}} - \sqrt{\left(\frac{v^2}{c_0^2}t'\right)^2 - 2\frac{v^2}{c_0^2}t'\sqrt{\frac{v^2}{c_0^2}t'^2 + \frac{y_0^2}{c_0^2}} + \frac{v^2}{c_0^2}t'^2 + \frac{y_0^2}{c_0^2}}\right)$$

$$\tau' = \frac{c_0^2}{c_0^2 - v^2}\left(t' - \sqrt{\frac{v^2}{c_0^2}t'^2 + \frac{y_0^2}{c_0^2}} - \sqrt{\left(\sqrt{\frac{v^2}{c_0^2}t'^2 + \frac{y_0^2}{c_0^2}} - \frac{v^2}{c_0^2}t'\right)^2}\right)$$

$$\tau' = \frac{c_0^2}{c_0^2 - v^2}\left(\left(1 + \frac{v^2}{c_0^2}\right)t' - 2\sqrt{\frac{v^2}{c_0^2}t'^2 + \frac{y_0^2}{c_0^2}}\right)$$

A.5 Derivation of Equation 5.25

$$v^2\tau^2 + y_0^2 = \left(c_0 t_r - \sqrt{v^2 t_r^2 + y_0^2} - c_0\tau\right)^2$$

$$\tau^2 - 2\frac{c_0 t_r - \sqrt{v^2 t_r^2 + y_0^2}}{c_0^2 - v^2}c_0\tau + \frac{\left(c_0 t_r - \sqrt{v^2 t_r^2 + y_0^2}\right)^2 - y_0^2}{c_0^2 - v^2} = 0$$

$$\tau = \frac{c_0 t_r - \sqrt{v^2 t_r^2 + y_0^2}}{c_0^2 - v^2}c_0 - \sqrt{\left(\frac{c_0 t_r - \sqrt{v^2 t_r^2 + y_0^2}}{c_0^2 - v^2}c_0\right)^2 - \frac{\left(c_0 t_r - \sqrt{v^2 t_r^2 + y_0^2}\right)^2 - y_0^2}{c_0^2 - v^2}}$$

$$\tau = \frac{c_0 t_r - \sqrt{v^2 t_r^2 + y_0^2}}{c_0^2 - v^2}c_0 - \sqrt{\cancel{\left(\frac{c_0 t_r - \sqrt{v^2 t_r^2 + y_0^2}}{c_0^2 - v^2}c_0\right)^2} - \frac{\cancel{c_0^2} - v^2}{c_0^2 - v^2}\frac{\left(c_0 t_r - \sqrt{v^2 t_r^2 + y_0^2}\right)^2}{c_0^2 - v^2} + \frac{y_0^2}{c_0^2 - v^2}}$$

$$\tau = \frac{c_0^2}{c_0^2 - v^2}\left(t_r - \sqrt{\frac{v^2 t_r^2}{c_0^2} + \frac{y_0^2}{c_0^2}} - \sqrt{\frac{v^2}{c_0^2}t_r^2 - 2\frac{v^2}{c_0^2}t_r\sqrt{\frac{v^2 t_r^2}{c_0^2} + \frac{y_0^2}{c_0^2}} + \frac{v^2}{c_0^2}\left(\frac{v^2 t_r^2}{c_0^2} + \cancel{\frac{y_0^2}{c_0^2}}\right) + \frac{c_0^2 \cancel{\times} y_0^2}{c_0^2\, c_0^2}}\right)$$

$$\tau = \frac{c_0^2}{c_0^2 - v^2}\left(t_r - \sqrt{\frac{v^2 t_r^2}{c_0^2} + \frac{y_0^2}{c_0^2}} - \sqrt{\frac{v^4}{c_0^4}t_r^2 - 2\frac{v^2}{c_0^2}t_r\sqrt{\frac{v^2 t_r^2}{c_0^2} + \frac{y_0^2}{c_0^2}} + \frac{v^2}{c_0^2}t_r^2 + \frac{y_0^2}{c_0^2}}\right)$$

$$\tau = \frac{c_0^2}{c_0^2 - v^2}\left(t_r - \sqrt{\frac{v^2 t_r^2}{c_0^2} + \frac{y_0^2}{c_0^2}} - \left(\sqrt{\frac{v^2 t_r^2}{c_0^2} + \frac{y_0^2}{c_0^2}} - \frac{v^2}{c_0^2}t_r\right)\right)$$

$$\tau = \frac{c_0^2}{c_0^2 - v^2}\left(\left(1 + \frac{v^2}{c_0^2}\right)t_r - 2\sqrt{\frac{v^2}{c_0^2}t_r^2 + \frac{y_0^2}{c_0^2}}\right)$$

A.6 Derivation of Equations 5.33a and 5.33b

$$c_0\tau + \sqrt{(x_s - v\tau)^2 + y_s^2 + h^2} = c_0 t - \sqrt{(x_s - vt)^2 + y_s^2 + h^2}$$

$$(x_s - v\tau)^2 + y_s^2 + h^2 = \left(c_0 t - \sqrt{(x_s - vt)^2 + y_s^2 + h^2} - c_0\tau\right)^2$$

$$c_0^2\tau^2 - (x_s - v\tau)^2 - 2c_0\tau\left(c_0 t - \sqrt{(x_s - vt)^2 + y_s^2 + h^2}\right) + \left(c_0 t - \sqrt{(x_s - vt)^2 + y_s^2 + h^2}\right)^2 - y_s^2 - h^2 = 0$$

$$\tau^2 - 2\frac{c_0^2 t - c_0\sqrt{(x_s - vt)^2 + y_s^2 + h^2} - x_s v}{c_0^2 - v^2}\tau + \frac{\left(c_0 t - \sqrt{(x_s - vt)^2 + y_s^2 + h^2}\right)^2 - x_s^2 - y_s^2 - h^2}{c_0^2 - v^2} = 0$$

$$\tau = \frac{c_0^2 t - c_0\sqrt{(x_s - vt)^2 + y_s^2 + h^2} - x_s v}{c_0^2 - v^2} - \sqrt{\left(\frac{c_0^2 t - c_0\sqrt{(x_s - vt)^2 + y_s^2 + h^2} - x_s v}{c_0^2 - v^2}\right)^2 - \frac{\left(c_0 t - \sqrt{(x_s - vt)^2 + y_s^2 + h^2}\right)^2 - x_s^2 - y_s^2 - h^2}{c_0^2 - v^2}}$$

$$\frac{d\tau}{dt} = \frac{c_0^2 - c_0\frac{d}{dt}\sqrt{(x_s - vt)^2 + y_s^2 + h^2}}{c_0^2 - v^2} - \frac{d}{dt}\sqrt{\left(\frac{c_0^2 t - c_0\sqrt{(x_s - vt)^2 + y_s^2 + h^2} - x_s v}{c_0^2 - v^2}\right)^2 - \frac{\left(c_0 t - \sqrt{(x_s - vt)^2 + y_s^2 + h^2}\right)^2 - x_s^2 - y_s^2 - h^2}{c_0^2 - v^2}}$$

$$\frac{d\tau}{dt} = \frac{c_0^2 - c_0\frac{d}{dt}\sqrt{(x_s - vt)^2 + y_s^2 + h^2}}{c_0^2 - v^2} - \frac{\frac{d}{dt}\left(\frac{c_0^2 t - c_0\sqrt{(x_s - vt)^2 + y_s^2 + h^2} - x_s v}{c_0^2 - v^2}\right)^2 - \frac{d}{dt}\frac{\left(c_0 t - \sqrt{(x_s - vt)^2 + y_s^2 + h^2}\right)^2 - x_s^2 - y_s^2 - h^2}{c_0^2 - v^2}}{2\sqrt{\left(\frac{c_0^2 t - c_0\sqrt{(x_s - vt)^2 + y_s^2 + h^2} - x_s v}{c_0^2 - v^2}\right)^2 - \frac{\left(c_0 t - \sqrt{(x_s - vt)^2 + y_s^2 + h^2}\right)^2 - x_s^2 - y_s^2 - h^2}{c_0^2 - v^2}}}$$

$$\frac{d\tau}{dt} = \frac{c_0^2 - c_0 \frac{d}{dt}\sqrt{(x_s - vt)^2 + y_s^2 + h^2}}{c_0^2 - v^2}$$

$$- \frac{\frac{c_0^2 t - c_0\sqrt{(x_s - vt)^2 + y_s^2 + h^2} - x_s v}{c_0^2 - v^2} \frac{d}{dt} \frac{c_0^2 t - c_0\sqrt{(x_s - vt)^2 + y_s^2 + h^2} - x_s v}{c_0^2 - v^2}}{\sqrt{\left(\frac{c_0^2 t - c_0\sqrt{(x_s - vt)^2 + y_s^2 + h^2} - x_s v}{c_0^2 - v^2}\right)^2 - \frac{\left(c_0 t - \sqrt{(x_s - vt)^2 + y_s^2 + h^2}\right)^2 - x_s^2 - y_s^2 - h^2}{c_0^2 - v^2}}}$$

$$+ \frac{\frac{\left(c_0 t - \sqrt{(x_s - vt)^2 + y_s^2 + h^2}\right)\left(c_0 - \frac{d}{dt}\sqrt{(x_s - vt)^2 + y_s^2 + h^2}\right)}{c_0^2 - v^2}}{\sqrt{\left(\frac{c_0^2 t - c_0\sqrt{(x_s - vt)^2 + y_s^2 + h^2} - x_s v}{c_0^2 - v^2}\right)^2 - \frac{\left(c_0 t - \sqrt{(x_s - vt)^2 + y_s^2 + h^2}\right)^2 - x_s^2 - y_s^2 - h^2}{c_0^2 - v^2}}}$$

$$\frac{d\tau}{dt} = \frac{c_0^2 - c_0 \frac{d}{dt}\sqrt{(x_s - vt)^2 + y_s^2 + h^2}}{c_0^2 - v^2} - \left(c_0 - \frac{d}{dt}\sqrt{(x_s - vt)^2 + y_s^2 + h^2}\right)$$

$$\cdot \frac{\frac{c_0^2 t - c_0\sqrt{(x_s - vt)^2 + y_s^2 + h^2} - x_s v}{c_0^2 - v^2} \frac{c_0}{c_0^2 - v^2} - \frac{\left(c_0 t - \sqrt{(x_s - vt)^2 + y_s^2 + h^2}\right)}{c_0^2 - v^2}}{\sqrt{\left(\frac{c_0^2 t - c_0\sqrt{(x_s - vt)^2 + y_s^2 + h^2} - x_s v}{c_0^2 - v^2}\right)^2 - \frac{\left(c_0 t - \sqrt{(x_s - vt)^2 + y_s^2 + h^2}\right)^2 - x_s^2 - y_s^2 - h^2}{c_0^2 - v^2}}}$$

$$\frac{d\tau}{dt} = \frac{c_0^2 - c_0 \frac{d}{dt}\sqrt{(x_s - vt)^2 + y_s^2 + h^2}}{c_0^2 - v^2} - \frac{c_0 - \frac{d}{dt}\sqrt{(x_s - vt)^2 + y_s^2 + h^2}}{c_0^2 - v^2}$$

$$\cdot \frac{\left(\cancel{c_0 t - \sqrt{(x_s - vt)^2 + y_s^2 + h^2}} - \frac{x_s v}{c_0}\right)\frac{c_0^2}{c_0^2 - v^2} - \frac{\cancel{c_0^2} - v^2}{c_0^2 - v^2}\left(c_0 t - \sqrt{(x_s - vt)^2 + y_s^2 + h^2}\right)}{\sqrt{\left(\frac{c_0^2 t - c_0\sqrt{(x_s - vt)^2 + y_s^2 + h^2} - x_s v}{c_0^2 - v^2}\right)^2 - \frac{\left(c_0 t - \sqrt{(x_s - vt)^2 + y_s^2 + h^2}\right)^2 - x_s^2 - y_s^2 - h^2}{c_0^2 - v^2}}}$$

$$\frac{d\tau}{dt} = \frac{c_0}{c_0^2 - v^2}\left(c_0 - \frac{d}{dt}\sqrt{(x_s - vt)^2 + y_s^2 + h^2}\right) +$$

$$\frac{\left(\frac{x_s v}{c_0}\frac{c_0^2}{c_0^2 - v^2} - \frac{v^2}{c_0^2 - v^2}\left(c_0 t - \sqrt{(x_s - vt)^2 + y_s^2 + h^2}\right)\right)\left(c_0 - \frac{d}{dt}\sqrt{(x_s - vt)^2 + y_s^2 + h^2}\right)}{\sqrt{\left(c_0^2 t - c_0\sqrt{(x_s - vt)^2 + y_s^2 + h^2} - x_s v\right)^2 - (c_0^2 - v^2)\left(\left(c_0 t - \sqrt{(x_s - vt)^2 + y_s^2 + h^2}\right)^2 - x_s^2 - y_s^2 - h^2\right)}}$$

$$\frac{d\tau}{dt} = \frac{c_0^2}{c_0^2 - v^2}\left(1 + \frac{v}{c_0}\frac{x_s - vt}{\sqrt{(x_s - vt)^2 + y_s^2 + h^2}}\right) +$$

$$\frac{\frac{c_0^2}{c_0^2 - v^2}\left(1 + \frac{v}{c_0}\frac{x_s - vt}{\sqrt{(x_s - vt)^2 + y_s^2 + h^2}}\right)\frac{v}{c_0}\left(c_0 x_s - v\left(c_0 t - \sqrt{(x_s - vt)^2 + y_s^2 + h^2}\right)\right)}{\sqrt{\left(c_0^2 t - c_0\sqrt{(x_s - vt)^2 + y_s^2 + h^2} - x_s v\right)^2 - (c_0^2 - v^2)\left(\left(c_0 t - \sqrt{(x_s - vt)^2 + y_s^2 + h^2}\right)^2 - x_s^2 - y_s^2 - h^2\right)}}$$

A.7 Taylor Expansion for Synthetic Aperture Radar

The point of expansion t_0 is calculated using eq. 5.33b.

$$1 - \frac{c_0^2}{c_0^2 - v^2}\left(1 + \frac{v}{c_0}\frac{x_s - vt_0}{\sqrt{(x_s - vt_0)^2 + y_s^2 + h^2}}\right) =$$

$$\frac{\frac{c_0^2}{c_0^2 - v^2}\left(1 + \frac{v}{c_0}\frac{x_s - vt_0}{\sqrt{(x_s - vt_0)^2 + y_s^2 + h^2}}\right)\frac{v}{c_0}\left(c_0 x_s - v\left(c_0 t_0 - \sqrt{(x_s - vt_0)^2 + y_s^2 + h^2}\right)\right)}{\sqrt{\left(c_0^2 t_0 - c_0\sqrt{(x_s - vt_0)^2 + y_s^2 + h^2} - x_s v\right)^2 - (c_0^2 - v^2)\left(\left(c_0 t_0 - \sqrt{(x_s - vt_0)^2 + y_s^2 + h^2}\right)^2 - x_s^2 - y_s^2 - h^2\right)}}$$

$$\frac{\left(\cancel{\frac{-v^2}{c_0^2}}\cancel{\frac{c_0^2}{c_0^2}} - \frac{v}{c_0}\frac{x_s - vt_0}{\sqrt{(x_s - vt_0)^2 + y_s^2 + h^2}}\right)}{\sqrt{\left(c_0^2 t_0 - c_0\sqrt{(x_s - vt_0)^2 + y_s^2 + h^2} - x_s v\right)^2 - (c_0^2 - v^2)\left(\left(c_0 t_0 - \sqrt{(x_s - vt_0)^2 + y_s^2 + h^2}\right)^2 - x_s^2 - y_s^2 - h^2\right)}}$$

$$= \frac{\left(1 + \frac{v}{c_0}\frac{x_s - vt_0}{\sqrt{(x_s - vt_0)^2 + y_s^2 + h^2}}\right)\frac{v}{c_0}\left(c_0 x_s - v\left(c_0 t_0 - \sqrt{(x_s - vt_0)^2 + y_s^2 + h^2}\right)\right)}{}$$

$$\frac{-\left(\cancel{\frac{v}{c_0^2}} + \cancel{\frac{v}{c_0}}\frac{x_s - vt_0}{\sqrt{(x_s - vt_0)^2 + y_s^2 + h^2}}\right).}{\sqrt{\left(c_0^2 t_0 - c_0\sqrt{(x_s - vt_0)^2 + y_s^2 + h^2} - x_s v\right)^2 - (c_0^2 - v^2)\left(\left(c_0 t_0 - \sqrt{(x_s - vt_0)^2 + y_s^2 + h^2}\right)^2 - x_s^2 - y_s^2 - h^2\right)}}$$

$$= \frac{\left(1 + \frac{v}{c_0}\frac{x_s - vt_0}{\sqrt{(x_s - vt_0)^2 + y_s^2 + h^2}}\right)\cancel{\frac{v}{c_0}}\left(c_0 x_s - v\left(c_0 t_0 - \sqrt{(x_s - vt_0)^2 + y_s^2 + h^2}\right)\right)}{}$$

$$\frac{-\cancel{\left(\frac{v}{c_0}\sqrt{(x_s - vt_0)^2 + y_s^2 + h^2} + x_s - vt_0\right)}.}{\sqrt{\left(c_0^2 t_0 - c_0\sqrt{(x_s - vt_0)^2 + y_s^2 + h^2} - x_s v\right)^2 - (c_0^2 - v^2)\left(\left(c_0 t_0 - \sqrt{(x_s - vt_0)^2 + y_s^2 + h^2}\right)^2 - x_s^2 - y_s^2 - h^2\right)}}$$

$$= \left(\sqrt{(x_s - vt_0)^2 + y_s^2 + h^2} + \frac{v}{c_0}(x_s - vt_0)\right)\left[c_0 x_s - vt_0 + \cancel{\frac{v}{c_0}\sqrt{(x_s - vt_0)^2 + y_s^2 + h^2}}\right]$$

$$\left(c_0^2 t_0 - c_0\sqrt{(x_s - vt_0)^2 + y_s^2 + h^2} - x_s v\right)^2 - \left(c_0^2 - v^2\right)\left(\left(c_0 t_0 - \sqrt{(x_s - vt_0)^2 + y_s^2 + h^2}\right)^2 - x_s^2 - y_s^2 - h^2\right)$$

$$= c_0^2 \left(\sqrt{(x_s - vt_0)^2 + y_s^2 + h^2} + \frac{v}{c_0}(x_s - vt_0)\right)^2$$

$$c_0^2\left(c_0 t_0 - \sqrt{(x_s - vt_0)^2 + y_s^2 + h^2} - x_s\frac{v}{c_0}\right)^2 - \left(c_0^2 - v^2\right)\left(c_0 t_0 - \sqrt{(x_s - vt_0)^2 + y_s^2 + h^2}\right)^2$$

$$+ \left(c_0^2 - v^2\right)\left(x_s^2 + y_s^2 + h^2\right) = c_0^2\left(\sqrt{(x_s - vt_0)^2 + y_s^2 + h^2} + \frac{v}{c_0}(x_s - vt_0)\right)^2$$

$$\cancel{c_0^2\left(c_0 t_0 - \sqrt{(x_s - vt_0)^2 + y_s^2 + h^2}\right)^2} - 2x_s v c_0\left(c_0 t_0 - \sqrt{(x_s - vt_0)^2 + y_s^2 + h^2}\right)$$

$$+ c_0^2\left(x_s \frac{v}{c_0}\right)^2 - \left(\cancel{c_0^2} - v^2\right)\left(c_0 t_0 - \sqrt{(x_s - vt_0)^2 + y_s^2 + h^2}\right)^2 + \left(c_0^2 - v^2\right)\left(x_s^2 + y_s^2 + h^2\right)$$

$$= c_0^2\left(\sqrt{(x_s - vt_0)^2 + y_s^2 + h^2} + \frac{v}{c_0}(x_s - vt_0)\right)^2$$

$$-2x_s v c_0\left(c_0 t_0 - \sqrt{(x_s - vt_0)^2 + y_s^2 + h^2}\right) + c_0^2\cancel{\left(x_s\frac{v}{c_0}\right)^2}$$

$$+ v^2\left(c_0 t_0 - \sqrt{(x_s - vt_0)^2 + y_s^2 + h^2}\right)^2 + c_0^2\left(x_s^2 + y_s^2 + h^2\right) - v^2\left(\cancel{x_s^2} + y_s^2 + h^2\right)$$

$$= c_0^2\left(\sqrt{(x_s - vt_0)^2 + y_s^2 + h^2} + \frac{v}{c_0}(x_s - vt_0)\right)^2$$

$$-2x_s v c_0\left(c_0 t_0 - \cancel{\sqrt{(x_s - vt_0)^2 + y_s^2 + h^2}}\right) + v^2\left(c_0 t_0 - \sqrt{(x_s - vt_0)^2 + y_s^2 + h^2}\right)^2 + c_0^2\left(x_s^2 + \cancel{y_s^2 + h^2}\right) -$$

$$v^2\left(y_s^2 + h^2\right) = c_0^2\left((x_s - vt_0)^2 + \cancel{y_s^2 + h^2}\right) + 2\frac{v}{c_0}(\cancel{x_s} - vt_0)c_0^2\left(\sqrt{(x_s - vt_0)^2 + y_s^2 + h^2}\right) + c_0^2\left(\frac{v}{c_0}(x_s - vt_0)\right)^2$$

$$-2x_s v c_0^2 t_0 + v^2\left(c_0 t_0 - \sqrt{(x_s - vt_0)^2 + y_s^2 + h^2}\right)^2 + c_0^2 x_s^2 - v^2\left(y_s^2 + h^2\right)$$

$$= \left(c_0^2 + v^2\right)(x_s - vt_0)^2 - 2v^2 c_0 t_0\left(\sqrt{(x_s - vt_0)^2 + y_s^2 + h^2}\right)$$

$$-2x_s vc_0^2 t_0 + v^2(c_0 t_0)^2 \;\cancel{-2v^2 c_0 t_0 \sqrt{(x_s-vt_0)^2+y_s^2+h^2}} + v^2\left((x_s-vt_0)^2+y_s^2+h^2\right)$$
$$+c_0^2 x_s^2 - v^2(y_s^2+h^2) = (c_0^2+\cancel{v^2})(x_s-vt_0)^2 \;\cancel{-2v^2 c_0 t_0 \sqrt{(x_s-vt_0)^2+y_s^2+h^2}}$$

$$-2x_s vc_0^2 t_0 + v^2(c_0 t_0)^2 + \cancel{v^2(y_s^2+h^2)} + c_0^2 x_s^2 \;\cancel{-v^2(y_s^2+h^2)} = c_0^2(x_s-vt_0)^2$$

$$\cancel{c_0^2}(x_s^2 - 2x_s vt_0 + v^2 t_0^2) = \cancel{c_0^2}(x_s-vt_0)^2 \Rightarrow x_s - vt_0 = \pm(x_s - vt_0)$$

The correct solution is $\quad x_s - vt_0 = vt_0 - x_s \Rightarrow \underline{\underline{t_0 = \frac{x_s}{v}}}$

Using $t_0 = x_s/v$ the coefficient a of the Taylor expansion according to eq. 5.35a can be calculated.

$$\tau(t_0) - t_0 = \frac{c_0^2 \frac{x_s}{v} - c_0\sqrt{\left(x_s - \cancel{\frac{x_s}{v}}\right)^2 + y_s^2 + h^2} - x_s v}{c_0^2 - v^2} - \frac{x_s}{v} -$$

$$\sqrt{\left(\frac{c_0^2 \frac{x_s}{v} - c_0\sqrt{\left(x_s - \cancel{\frac{x_s}{v}}\right)^2 + y_s^2 + h^2} - x_s v}{c_0^2 - v^2}\right)^2 - \frac{\left(c_0 \frac{x_s}{v} - \sqrt{\left(x_s - \cancel{\frac{x_s}{v}}\right)^2 + y_s^2 + h^2}\right)^2 - x_s^2 - y_s^2 - h^2}{c_0^2 - v^2}}$$

$$\tau(t_0) - t_0 = \frac{\cancel{c_0^2 \frac{x_s}{v}} - c_0\sqrt{y_s^2+h^2} \;\cancel{-x_s v}}{c_0^2 - v^2} - \frac{\cancel{}}{c_0^2 - v^2}\frac{x_s}{v}$$

$$- \sqrt{\left(\frac{c_0^2 \frac{x_s}{v} - c_0\sqrt{y_s^2+h^2} - x_s v}{c_0^2 - v^2}\right)^2 - \frac{\left(c_0 \frac{x_s}{v} - \sqrt{y_s^2+h^2}\right)^2 - x_s^2 - y_s^2 - h^2}{c_0^2 - v^2}}$$

$$\tau(t_0) - t_0 = -\frac{c_0\sqrt{y_s^2+h^2}}{c_0^2 - v^2} - \sqrt{\left(\frac{c_0^2 \frac{x_s}{v} - c_0\sqrt{y_s^2+h^2} - x_s v}{c_0^2 - v^2}\right)^2 - \frac{\left(c_0 \frac{x_s}{v} - \sqrt{y_s^2+h^2}\right)^2 - x_s^2 - y_s^2 - h^2}{c_0^2 - v^2}}$$

$$\tau(t_0) - t_0 = -\frac{c_0}{c_0^2 - v^2}\sqrt{y_s^2 + h^2}$$

$$-\frac{1}{c_0^2 - v^2}\sqrt{\left(c_0^2\frac{x_s}{v} - c_0\sqrt{y_s^2 + h^2} - x_s v\right)^2 - (c_0^2 - v^2)\left(\left(c_0\frac{x_s}{v} - \sqrt{y_s^2 + h^2}\right)^2 - x_s^2 - y_s^2 - h^2\right)}$$

$$\tau(t_0) - t_0 = -\frac{c_0}{c_0^2 - v^2}\sqrt{y_s^2 + h^2}$$

$$-\frac{1}{c_0^2 - v^2}\sqrt{c_0^2\left(c_0\frac{x_s}{v} - \sqrt{y_s^2 + h^2} - x_s\frac{v}{c_0}\right)^2 - (c_0^2 - v^2)\left(\left(c_0\frac{x_s}{v} - \sqrt{y_s^2 + h^2}\right)^2 - x_s^2 - y_s^2 - h^2\right)}$$

$$\tau(t_0) - t_0 = -\frac{c_0}{c_0^2 - v^2}\sqrt{y_s^2 + h^2}$$

$$-\frac{1}{c_0^2 - v^2}\left(c_0^2\left(c_0\frac{x_s}{v} - \sqrt{y_s^2 + h^2}\right)^2 - 2x_s\frac{v}{c_0}c_0^2\left(c_0\frac{x_s}{v} - \sqrt{y_s^2 + h^2}\right) + c_0^2\left(x_s\frac{v}{c_0}\right)^2\right.$$
$$\left. - c_0^2\left(\left(c_0\frac{x_s}{v} - \sqrt{y_s^2 + h^2}\right)^2 - x_s^2 - y_s^2 - h^2\right) + v^2\left(\left(c_0\frac{x_s}{v} - \sqrt{y_s^2 + h^2}\right)^2 - x_s^2 - y_s^2 - h^2\right)\right)^{\frac{1}{2}}$$

$$\tau(t_0) - t_0 = -\frac{c_0}{c_0^2 - v^2}\sqrt{y_s^2 + h^2}$$

$$-\frac{1}{c_0^2 - v^2}\sqrt{\underbrace{v^2\left(c_0\frac{x_s}{v} - \sqrt{y_s^2 + h^2}\right)^2 - 2x_s v c_0\left(c_0\frac{x_s}{v} - \sqrt{y_s^2 + h^2}\right) + c_0^2 x_s^2}_{\left(v\left(c_0\frac{x_s}{v} - \sqrt{y_s^2+h^2}\right) - c_0 x_s\right)^2} + c_0^2(y_s^2 + h^2) - v^2(y_s^2 + h^2)}$$

$$\tau(t_0) - t_0 = -\frac{c_0}{c_0^2 - v^2}\sqrt{y_s^2 + h^2} - \frac{1}{c_0^2 - v^2}\sqrt{\left(v\left(c_0\frac{x_s}{v} - \sqrt{y_s^2 + h^2}\right) - c_0 x_s\right)^2 + (c_0^2 - v^2)(y_s^2 + h^2)}$$

$$\tau(t_0) - t_0 = -\frac{c_0}{c_0^2 - v^2}\sqrt{y_s^2 + h^2} - \frac{1}{c_0^2 - v^2}\sqrt{v^2(y_s^2 + h^2) + (c_0^2 - v^2)(y_s^2 + h^2)}$$

$$\tau(t_0) - t_0 = -\frac{c_0}{c_0^2 - v^2}\sqrt{y_s^2 + h^2} - \frac{c_0}{c_0^2 - v^2}\sqrt{y_s^2 + h^2} \Rightarrow \underline{\underline{a = -4\pi f_0 \frac{c_0}{c_0^2 - v^2}\sqrt{y_s^2 + h^2}}}$$

For the calculation of the coefficient c of the Taylor expansion according to eq. 5.35a the calculation of the second derivative of the phase difference $\Delta\varphi(t)$ resp. the transmission time $\tau(t)$ is required.

$$\frac{d^2\tau}{dt^2} = \frac{d}{dt}\frac{c_0^2}{c_0^2-v^2}\left(1+\frac{v}{c_0}\frac{x_s-vt}{\sqrt{(x_s-vt)^2+y_s^2+h^2}}\right) +$$

$$\frac{d}{dt}\frac{\frac{c_0^2}{c_0^2-v^2}\left(1+\frac{v}{c_0}\frac{x_s-vt}{\sqrt{(x_s-vt)^2+y_s^2+h^2}}\right)\frac{v}{c_0}\left(c_0x_s-v\left(c_0t-\sqrt{(x_s-vt)^2+y_s^2+h^2}\right)\right)}{\sqrt{\left(c_0^2t-c_0\sqrt{(x_s-vt)^2+y_s^2+h^2}-x_sv\right)^2-(c_0^2-v^2)\left(\left(c_0t-\sqrt{(x_s-vt)^2+y_s^2+h^2}\right)^2-x_s^2-y_s^2-h^2\right)}}$$

$$\frac{d^2\tau}{dt^2} = \frac{d}{dt}\frac{c_0^2}{c_0^2-v^2}\left(1+\frac{v}{c_0}\frac{x_s-vt}{\sqrt{(x_s-vt)^2+y_s^2+h^2}}\right) +$$

$$\left(\left(c_0^2t-c_0\underbrace{\sqrt{(x_s-vt)^2+y_s^2+h^2}}_{=0 \text{ for } t=x_s/v}-x_sv\right)^2 -(c_0^2-v^2)\left(\left(c_0t-\underbrace{\sqrt{(x_s-vt)^2+y_s^2+h^2}}_{=0 \text{ for } t=x_s/v}\right)^2-x_s^2-y_s^2-h^2\right)\right)^{-\frac{1}{2}}$$

$$\cdot\frac{d}{dt}\frac{c_0^2}{c_0^2-v^2}\left(1+\frac{v}{c_0}\frac{x_s-vt}{\sqrt{(x_s-vt)^2+y_s^2+h^2}}\right)\frac{v}{c_0}\left(c_0x_s-v\left(c_0t-\sqrt{(x_s-vt)^2+y_s^2+h^2}\right)\right)$$

$$+\frac{c_0^2}{c_0^2-v^2}\left(1+\frac{v}{c_0}\underbrace{\frac{x_s-vt}{\sqrt{(x_s-vt)^2+y_s^2+h^2}}}_{=0 \text{ for } t=x_s/v}\right)\frac{v}{c_0}\left(c_0x_s-v\left(c_0t-\underbrace{\sqrt{(x_s-vt)^2+y_s^2+h^2}}_{=0 \text{ for } t=x_s/v}\right)\right)\cdot$$

$$\frac{d}{dt}\left(\left(c_0^2t-c_0\sqrt{(x_s-vt)^2+y_s^2+h^2}-x_sv\right)^2-(c_0^2-v^2)\left(\left(c_0t-\sqrt{(x_s-vt)^2+y_s^2+h^2}\right)^2-x_s^2-y_s^2-h^2\right)\right)^{-\frac{1}{2}}$$

$$\frac{d^2\tau}{dt^2} = \frac{d}{dt}\frac{c_0^2}{c_0^2-v^2}\left(1+\frac{v}{c_0}\frac{x_s-vt}{\sqrt{(x_s-vt)^2+y_s^2+h^2}}\right)+$$

$$\left(\left(c_0^2 t-c_0\sqrt{y_s^2+h^2}-x_s v\right)^2-\left(c_0^2-v^2\right)\left(\left(c_0 t-\sqrt{y_s^2+h^2}\right)^2-x_s^2-y_s^2-h^2\right)\right)^{-\frac{1}{2}}$$

$$\cdot\frac{d}{dt}\frac{c_0^2}{c_0^2-v^2}\left(1+\frac{v}{c_0}\frac{x_s-vt}{\sqrt{(x_s-vt)^2+y_s^2+h^2}}\right)\frac{v}{c_0}\left(c_0 x_s-v\left(c_0 t-\sqrt{(x_s-vt)^2+y_s^2+h^2}\right)\right)$$

$$-\frac{1}{2}\frac{c_0^2}{c_0^2-v^2}\frac{v}{c_0}\left(c_0 x_s-v\left(c_0 t-\sqrt{y_s^2+h^2}\right)\right)\cdot$$

$$\left(\left(c_0^2 t-c_0\underbrace{\sqrt{(x_s-vt)^2+y_s^2+h^2}}_{=0 \text{ for } t=x_s/v}-x_s v\right)^2-\left(c_0^2-v^2\right)\left(\left(c_0 t-\underbrace{\sqrt{(x_s-vt)^2+y_s^2+h^2}}_{=0 \text{ for } t=x_s/v}\right)^2-x_s^2-y_s^2-h^2\right)\right)^{-\frac{3}{2}}$$

$$\cdot\left(\frac{d}{dt}\left(c_0^2 t-c_0\sqrt{(x_s-vt)^2+y_s^2+h^2}-x_s v\right)^2-\left(c_0^2-v^2\right)\frac{d}{dt}\left(c_0 t-\sqrt{(x_s-vt)^2+y_s^2+h^2}\right)^2\right)$$

Auxiliary calculation 1:

$$\frac{d}{dt}\frac{c_0^2}{c_0^2-v^2}\left(1+\frac{v}{c_0}\frac{x_s-vt}{\sqrt{(x_s-vt)^2+y_s^2+h^2}}\right)\frac{v}{c_0}\left(c_0 x_s - v\left(c_0 t - \sqrt{(x_s-vt)^2+y_s^2+h^2}\right)\right)$$

$$= \frac{c_0^2}{c_0^2-v^2}\left(1+\frac{v}{c_0}\underbrace{\frac{x_s-vt}{\sqrt{(x_s-vt)^2+y_s^2+h^2}}}_{=0 \text{ for } t=x_s/v}\right)\frac{v}{c_0}\frac{d}{dt}\left(c_0 x_s - v\left(c_0 t - \sqrt{(x_s-vt)^2+y_s^2+h^2}\right)\right)$$

$$+\frac{v}{c_0}\left(\underbrace{c_0 x_s - vc_0 t}_{=0 \text{ for } t=x_s/v}+v\underbrace{\sqrt{(x_s-vt)^2+y_s^2+h^2}}_{=0 \text{ for } t=x_s/v}\right)\frac{c_0^2}{c_0^2-v^2}\frac{d}{dt}\left(1+\frac{v}{c_0}\frac{x_s-vt}{\sqrt{(x_s-vt)^2+y_s^2+h^2}}\right)$$

$$=\frac{v^2}{c_0}\sqrt{y_s^2+h^2}\frac{c_0^2}{c_0^2-v^2}\frac{d}{dt}\left(1+\frac{v}{c_0}\frac{x_s-vt}{\sqrt{(x_s-vt)^2+y_s^2+h^2}}\right) - \frac{c_0^2}{c_0^2-v^2}\frac{v^2}{c_0}\left(c_0-\frac{d}{dt}\sqrt{(x_s-vt)^2+y_s^2+h^2}\right)$$

$$=\frac{v^2}{c_0}\sqrt{y_s^2+h^2}\frac{c_0^2}{c_0^2-v^2}\frac{d}{dt}\left(1+\frac{v}{c_0}\frac{x_s-vt}{\sqrt{(x_s-vt)^2+y_s^2+h^2}}\right) - \frac{c_0^2}{c_0^2-v^2}\frac{v^2}{c_0}\left(c_0 - \underbrace{\frac{2(x_s-vt)\frac{d}{dt}(x_s-vt)}{2\sqrt{(x_s-vt)^2+y_s^2+h^2}}}_{=0 \text{ for } t=x_s/v}\right)$$

$$=-\frac{c_0^2 v^2}{c_0^2-v^2}+\sqrt{y_s^2+h^2}\frac{c_0 v^2}{c_0^2-v^2}\frac{d}{dt}\left(1+\frac{v}{c_0}\frac{x_s-vt}{\sqrt{(x_s-vt)^2+y_s^2+h^2}}\right)$$

Auxiliary calculation 2:

$$\frac{d}{dt}\left(c_0^2 t - c_0\sqrt{(x_s-vt)^2+y_s^2+h^2} - x_s v\right)^2 - (c_0^2-v^2)\frac{d}{dt}\left(c_0 t - \sqrt{(x_s-vt)^2+y_s^2+h^2}\right)^2$$

$$= 2\left(c_0^2 t - c_0\underbrace{\sqrt{(x_s-vt)^2+y_s^2+h^2}}_{=0 \text{ for } t=x_s/v} - x_s v\right)\frac{d}{dt}\left(c_0^2 t - c_0\sqrt{(x_s-vt)^2+y_s^2+h^2} - x_s v\right)$$

$$-(c_0^2-v^2)2\left(c_0 t - \underbrace{\sqrt{(x_s-vt)^2+y_s^2+h^2}}_{=0 \text{ for } t=x_s/v}\right)\frac{d}{dt}\left(c_0 t - \sqrt{(x_s-vt)^2+y_s^2+h^2}\right)$$

$$= \left(2\left(c_0^2 t - c_0\sqrt{y_s^2 + h^2} - x_s v\right)c_0 - \left(c_0^2 - v^2\right)2\left(c_0 t - \sqrt{y_s^2 + h^2}\right)\right)\left(c_0 - \underbrace{\frac{2(x_s - vt)\frac{d}{dt}(x_s - vt)}{2\sqrt{(x_s - vt)^2 + y_s^2 + h^2}}}_{=0 \text{ for } t=x_s/v}\right)$$

$$= 2\left(\cancel{c_0^2 t - c_0\sqrt{y_s^2 + h^2}} - x_s v\right)c_0^2 - \cancel{c_0^2 2\left(c_0 t - \sqrt{y_s^2 + h^2}\right)c_0} + v^2 2\left(c_0 t - \sqrt{y_s^2 + h^2}\right)c_0$$

$$= \underbrace{-2x_s vc_0^2 + v^2 2c_0^2 t}_{=0 \text{ for } t=x_s/v} - v^2 2c_0\sqrt{y_s^2 + h^2} = -2v^2 c_0 \sqrt{y_s^2 + h^2}$$

Auxiliary calculation 3:

$$\frac{d}{dt}\left(1 + \frac{v}{c_0}\frac{x_s - vt}{\sqrt{(x_s - vt)^2 + y_s^2 + h^2}}\right) = \frac{v}{c_0}\frac{d}{dt}\left((x_s - vt)^2 + y_s^2 + h^2\right)^{-\frac{1}{2}}(x_s - vt)$$

$$= \frac{v}{c_0}\left(\underbrace{\left[(x_s - vt)^2 + y_s^2 + h^2\right]^{-\frac{1}{2}}\frac{d}{dt}(x_s - vt)}_{=0 \text{ for } t=x_s/v} + \frac{v}{c_0}\underbrace{(x_s - vt)\frac{d}{dt}\left((x_s - vt)^2 + y_s^2 + h^2\right)^{-\frac{1}{2}}}_{=0 \text{ for } t=x_s/v}\right) = -\frac{v^2}{c_0\sqrt{y_s^2 + h^2}}$$

$$\frac{d^2\tau}{dt^2} = -\frac{v^2}{c_0}\frac{1}{\sqrt{y_s^2 + h^2}}\frac{c_0^2}{c_0^2 - v^2} -$$

$$\frac{v^2}{c_0^2 - v^2}\left(c_0^2 + v^2\right)\cdot\left(\left(c_0^2 t - c_0\sqrt{y_s^2 + h^2} - x_s v\right)^2 - \left(c_0^2 - v^2\right)\left(\left(c_0 t - \sqrt{y_s^2 + h^2}\right)^2 - x_s^2 - y_s^2 - h^2\right)\right)^{-\frac{1}{2}}$$

$$+ \frac{c_0^2 v^2}{c_0^2 - v^2} v\sqrt{y_s^2 + h^2}\left(\underbrace{c_0 x_s - vc_0 t}_{=0 \text{ for } t=x_s/v} + v\sqrt{y_s^2 + h^2}\right)$$

$$\cdot\left(\left(c_0^2 t - c_0\sqrt{y_s^2 + h^2} - x_s v\right)^2 - \left(c_0^2 - v^2\right)\left(\left(c_0 t - \sqrt{y_s^2 + h^2}\right)^2 - x_s^2 - y_s^2 - h^2\right)\right)^{-\frac{3}{2}}.$$

$$\frac{d^2\tau}{dt^2} = -\frac{v^2}{c_0}\frac{1}{\sqrt{y_s^2 + h^2}}\frac{c_0^2}{c_0^2 - v^2} -$$

$$\frac{v^2}{c_0^2 - v^2}\left(c_0^2 + v^2\right)\cdot\left(\left(c_0^2 t - c_0\sqrt{y_s^2 + h^2} - x_s v\right)^2 - \left(c_0^2 - v^2\right)\left(\left(c_0 t - \sqrt{y_s^2 + h^2}\right)^2 - x_s^2 - y_s^2 - h^2\right)\right)^{-\frac{1}{2}}$$

$$+ \frac{c_0^2 v^2}{c_0^2 - v^2} v^2\left(y_s^2 + h^2\right)\cdot\left(\left(c_0^2 t - c_0\sqrt{y_s^2 + h^2} - x_s v\right)^2 - \left(c_0^2 - v^2\right)\left(\left(c_0 t - \sqrt{y_s^2 + h^2}\right)^2 - x_s^2 - y_s^2 - h^2\right)\right)^{-\frac{3}{2}}.$$

Auxiliary calculation 4:

$$\left(c_0^2 t - c_0\sqrt{y_s^2+h^2} - x_s v\right)^2 - \left(c_0^2 - v^2\right)\left[\left(c_0 t - \sqrt{y_s^2+h^2}\right)^2 - x_s^2 - y_s^2 - h^2\right]$$

$$= c_0^2\left(c_0 t - \sqrt{y_s^2+h^2} - x_s\frac{v}{c_0}\right)^2 + \left(c_0^2 - v^2\right)\left(x_s^2 + y_s^2 + h^2 - \left(c_0 t - \sqrt{y_s^2+h^2}\right)^2\right)$$

$$= c_0^2\cancel{\left(c_0 t - \sqrt{y_s^2+h^2}\right)^2} - 2x_s v c_0\left(c_0 t - \sqrt{y_s^2+h^2}\right) + c_0^2\cancel{\left(x_s\frac{v}{c_0}\right)^2} + c_0^2\left(x_s^2 + y_s^2 + h^2 - \cancel{\left(c_0 t - \sqrt{y_s^2+h^2}\right)^2}\right)$$
$$- v^2\left(\cancel{x_s^2} + y_s^2 + h^2 - \left(c_0 t - \sqrt{y_s^2+h^2}\right)^2\right)$$

$$= c_0^2 x_s^2 - 2x_s v c_0\left(c_0 t - \sqrt{y_s^2+h^2}\right) + v^2\left(c_0 t - \sqrt{y_s^2+h^2}\right)^2 + c_0^2\left(y_s^2+h^2\right) - v^2\left(y_s^2+h^2\right)$$

$$= \Big(\underbrace{c_0 x_s - v c_0 t + v\sqrt{y_s^2+h^2}}_{=0 \text{ for } t=x_s/v}\Big)^2 + c_0^2\left(y_s^2+h^2\right) - v^2\left(y_s^2+h^2\right)$$

$$= \cancel{\left(v\sqrt{y_s^2+h^2}\right)^2} + c_0^2\left(y_s^2+h^2\right) \cancel{- v^2\left(y_s^2+h^2\right)} = c_0^2\left(y_s^2+h^2\right)$$

Combining the results of the auxiliary calculations the second derivative of the transmission time $\tau(t_0)$ at the point of expansion $t_0 = x_s/v$ is obtained.

$$\left.\frac{d^2\tau}{dt^2}\right|_{t=\frac{x_s}{v}} = -\frac{v^2}{c_0}\frac{1}{\sqrt{y_s^2+h^2}}\frac{c_0^2}{c_0^2-v^2} - \frac{v^2}{c_0^2-v^2}\left(c_0^2+v^2\right)\frac{1}{c_0\sqrt{y_s^2+h^2}} + \frac{c_0^2 v^2}{c_0^2-v^2}v^2\left(y_s^2+h^2\right)\left(\frac{1}{c_0\sqrt{y_s^2+h^2}}\right)^3$$

$$\left.\frac{d^2\tau}{dt^2}\right|_{t=\frac{x_s}{v}} = -\frac{v^2}{c_0}\frac{1}{\sqrt{y_s^2+h^2}}\frac{c_0^2}{c_0^2-v^2} - \frac{v^2}{c_0^2-v^2}\left(c_0^2+\cancel{v^2}\right)\frac{1}{c_0\sqrt{y_s^2+h^2}} + \cancel{\frac{v^2}{c_0^2-v^2}\frac{1}{c_0\sqrt{y_s^2+h^2}}}$$

$$\left.\frac{d^2\tau}{dt^2}\right|_{t=\frac{x_s}{v}} = -2\frac{v^2}{\sqrt{y_s^2+h^2}}\frac{c_0}{c_0^2-v^2} \Rightarrow \underline{\underline{c = -\frac{2\pi f_0}{\sqrt{y_s^2+h^2}}\frac{c_0 v^2}{c_0^2-v^2}}}$$

Acronyms and Symbols

ACF	Auto Correlation Function
SAR	Synthetic Aperture Radar
a, b, c, t_0	Taylor coefficients and point of expansion
c_0	velocity of light in vacuum ≈ velocity of light in air
d	constant distance between transmitter and receiver
d(t)	time-variant distance between transmitter and receiver
$\delta(t)$	Dirac's delta function
f	frequency
f(t)	instantaneous frequency of received signal as a function of time
f_0	constant frequency of sinusoidal transmitted signal
f_r	constant frequency of sinusoidal received signal
f_R	rotational speed of circulating transmitter
f_t	constant frequency of sinusoidal transmitted signal
g(t)	received signal
$\underline{G}(f)$	Fourier transform of received signal
h	flight altitude of an airborne radar over ground at z = 0
h(t - τ)	impulse response of a linear time-invariant system
h(t,τ)	impulse response of a linear time-variant system
R	orbit radius of circulating transmitter
$\vec{r} = (x, y, z)$	Cartesian coordinates
$\vec{r}_0$	fixed spatial position
$\vec{r}_{r0}$	spatial position of receiver at time t = 0
$\vec{r}_{t0}$	spatial position of transmitter at time t = 0
$\vec{r}_r(t)$	trajectory of receiver
$\dot{\vec{r}}_r(t)$	instantaneous velocity of receiver
$\ddot{\vec{r}}_r(t)$	instantaneous acceleration of receiver
$\vec{r}_t(t)$	trajectory of transmitter
$\dot{\vec{r}}_t(t)$	instantaneous velocity of transmitter
$\ddot{\vec{r}}_t(t)$	instantaneous acceleration of transmitter
s(t)	transmitted signal
$\underline{S}(f)$	Fourier transform of transmitted signal

t		temporal coordinate
t'		temporal coordinate after application of Lorentz transform
t_0		At this time the instantaneous frequency is equal to the frequency of the transmitted signal.
t_r		receiving time
t'_r		receiving time after application of Lorentz transform
t_s		scattering time
t'_s		scattering time after application of Lorentz transform
τ		transmission time
τ'		transmission time after application of Lorentz transform
$\hat{\tau}$		transmission time renamed for calculation of ACF
τ_0		point of expansion for linearisation of trajectory of transmitter
u, w		auxiliary variables for changes of variables
v_d		relativistic subtraction of velocities
v_r		velocity of receiver in uniform motion
v_t		velocity of transmitter in uniform motion
$\vec{v}$		velocity vector of radar resp. reflector in uniform motion
$\vec{v}_d$		relativistic subtraction of velocity vectors
$\vec{v}_r$		velocity vector of receiver in uniform motion
$\vec{v}_t$		velocity vector of transmitter in uniform motion
x, y, z		Cartesian coordinates
x_C		orbit centre of circulating transmitter
x_{r0}		position of receiver on the x-axis of the coordinate system at time t = 0
x_s, y_s		fixed position of reflector on the plane z = 0
x_{t0}		position of transmitter on the x-axis of the coordinate system at time t = 0
y_0		constant distance of transmitter resp. receiver to x-axis of the coordinate system
θ		angle between line of sight from the receiver to the transmitter and motion vector
$\varphi(t)$		instantaneous phase of received signal as a function of time
$\Delta\varphi(t)$		instantaneous phase difference as a function of time
$\varphi_{gg}(\Delta t)$		ACF of received signal at lag Δt
$\varphi_{ss}(\Delta t)$		ACF of transmitted signal at lag Δt
*		convolution operator
·		scalar (inner) product of two vectors

References

[1] C. Doppler: *Über das farbige Licht der Doppelsterne und einiger anderer Gestirne des Himmels*, in: *Abh. königl. böhm. Ges. Wiss.* **2**, Prague, 1843, 465-482

[2] A. Einstein: *Zur Elektrodynamik bewegter Körper*, in: *Annalen der Physik und Chemie* 17/1905, 891-921

[3] B. P. Lathi: *Linear Signals and Systems*, Oxford University Press, New York Oxford, 2002

[4] R. Sexl, H. K. Schmidt: *Raum Zeit Relativität*, 4. ed., Vieweg, Wiesbaden, Germany 2000

[5] A. Müller: *Elektromagnetische Schwingungen und Wellen, Wellenoptik, Relativititätstheorie*, 2. ed., Oldenbourg, München, Ehrenwirth, Germany, 1995

[6] W. Rindler: *Relativity, special, general and cosmological*, 2. ed., Oxford University Press, Oxford, 2006

[7] J. Freund: *Special Relativity for Beginners*, World Scientific Publishing Co., Singapore, 2008

[8] A. Einstein: *Die Grundlage der allgemeinen Relativitätstheorie*: in: *Annalen der Physik* 7/1916, 769-822

[9] I. G. Cumming, F. H. Wong: *Digital Processing of Synthetic Aperture Radar*, Artech House, Boston, Massachusetts, 2005

Index

acceleration, 57
acoustic, 5, 13, 40, 60-62
airborne, 49
amplitude scaling, 14, 18, 19, 21, 31, 46, 62
analytic calculation, 62
Auto Correlation Function (ACF), 15, 16, 30, 31, 62
bistatic, 62
causality condition, 12, 13, 24, 25, 34, 41, 42, 44, 48, 50, 55, 59
circular motion, 60
closed form, 33, 56, 57, 59
convolution, 10
coordinate system, 11, 17, 19, 22, 26, 28, 34, 37, 38, 40, 47, 66
Dirac's delta function, 10, 14-16, 18, 19, 25, 55-58
dispersion, 62
Doppler, Christian, 5, 9, 85
Doppler factor, 19, 22
Doppler scaling, 14
Doppler shift, 14
Einstein, Albert, 5, 9, 18, 36, 46, 62, 85
Fourier transform, 16, 32, 33
frequency scaling, 16, 27, 29, 35
frequency shift, 14
general relativity, 57, 62, 85
ground, 49
impulse response, 10, 14, 24, 33, 38, 42, 55, 57, 59
instantaneous frequency, 26, 27, 29, 35, 36, 38, 44, 46, 56, 58, 60-62
instantaneous velocity, 57, 59-61
light cone, 13
light, velocity of, 9, 20-22, 65
linear superposition, 10, 13, 55, 57
linearisation, 56, 57, 59
longitudinal, 26, 28, 29
Lorentz transform, 17, 18, 33, 37, 45, 49, 57, 59, 62
monostatic radar, 40, 62
multistatic, 62
non-uniform motion, 55, 57, 58, 62
normalised instantaneous frequency, 27, 29, 35, 37, 44, 47, 51, 52, 60, 61
orthogonal, 64, 65

parabolic, 52
phase, 27, 29, 35, 52,-54, 78
phase difference, 52-54, 78
point scatterer, 40
quadratic equation, 24, 25, 30, 34, 41, 43, 48, 50
radar, 5, 40-42, 45-50, 52, 62, 74, 85
receiving time, 17, 24, 25, 33, 43, 50, 57, 59
reflector, 5, 40-49, 52, 62
reflection coefficient, 49
relative velocity, 9
relativity, 5, 9, 17, 57, 62, 64, 85
remote sensing, 49
satellite based, 49
scalar product, 23,, 24, 64
scaling factor, 14, 19, 21, 22, 27, 29, 31, 35, 40
scaling property, 16
scattering time, 41, 42, 48-50
second derivative, 54, 78, 82
shifting property, 10
sifting property, 14-16, 18, 19, 25
signal amplitude, 14, 44-46, 51
signal delay, 10, 14, 16-19, 22, 62
sinusoidal waveform, 5, 9, 11, 14, 26, 28, 35, 44, 46, 52, 56, 58, 62
space-time diagram, 12, 13, 17
spatial position, function of, 36, 37, 44, 47, 52
special relativity, 5, 9, 17, 57, 62, 64, 85
sphere, 11, 23, 40, 41, 48-50, 55, 58
spherical wave, 11, 12, 23, 24, 40-41, 48-50, 55, 58
stochastic signal, 16, 31
subtraction of velocities, 9, 20-22, 38, 39, 64
supersonic, 14
surface of the earth, 49
Synthetic Aperture Radar (SAR), 52, 74, 85
Taylor expansion, 53, 54, 74, 76, 78
time delay, 14, 16-19, 21, 22, 62
time dependent, 10
time, function of, 11, 23, 33, 40, 41, 48-50, 55, 56, 58, 62
time scaling, 14, 16, 18, 33, 62
time variant, 24, 33, 38, 42, 55, 57, 59
trajectory, 12, 24, 41, 48, 50, 55, 58, 60, 61
transmission time, 18, 36, 37, 54, 59, 78, 82

transverse, 26, 36
unit vector, 65
velocity of propagation, 9, 11, 14, 23, 40, 48, 49, 55, 58, 62
velocity of sound, 9, 14
velocity vector, 64-66
worldline, 13

Die VDM Verlagsservicegesellschaft sucht für wissenschaftliche Verlage abgeschlossene und herausragende

Dissertationen, Habilitationen, Diplomarbeiten, Master Theses, Magisterarbeiten usw.

für die kostenlose Publikation als Fachbuch.

Sie verfügen über eine Arbeit, die hohen inhaltlichen und formalen Ansprüchen genügt, und haben Interesse an einer honorarvergüteten Publikation?

Dann senden Sie bitte erste Informationen über sich und Ihre Arbeit per Email an *info@vdm-vsg.de*.

Sie erhalten kurzfristig unser Feedback!

VDM Verlagsservicegesellschaft mbH
Dudweiler Landstr. 99   Telefon +49 681 3720 174
D - 66123 Saarbrücken   Fax   +49 681 3720 1749

www.vdm-vsg.de

Die VDM Verlagsservicegesellschaft mbH vertritt

Printed by Books on Demand GmbH, Norderstedt / Germany